再也不踩坑的
Kubernetes
实战指南

杜 宽 编著

清华大学出版社
北京

内 容 简 介

本书以实战为主线,深入浅出地介绍了 Kubernetes 在企业生产环境中的应用。全书共 6 章,第 1 章讲解 Kubernetes 的高可用安装,分为 kubeadm 和二进制安装方式,可以让读者快速上手,了解 Kubernetes 的架构模式。第 2 章介绍了 Kubernetes 的理论基础。第 3 章主要讲解 Kubernetes 常见应用的容器化,并部署至 Kubernetes 集群实现高可用,同时介绍了 Kubernetes 的各个组件和资源。第 4 章主要介绍持续集成和持续部署,包括 Jenkins 最新的功能 Pipeline 的使用,从 Pipeline 的语法到项目实操,传统 Java 和 Spring Cloud 应用的容器化以及自动化构建部署。第 5 章主要讲解了 Kubernetes 的 Nginx Ingress 的安装和常用配置,以适用于生产环境的各种需求。第 6 章讲解了备受关注的 Server Mesh,使用 Istio 代替微服务架构中的网络功能、实现限速、分流和路由等内容。

本书技术先进,注重实操,解决了 Kubernetes 在生产环境中使用和安装过程中遇到的大量问题,本书特别适合想尽快将 Kubernetes 应用于其公司业务中的 Kubernetes 初学者、开发人员、运维工程师和架构师使用。

本书封面贴有清华大学出版社防伪标签,无标签者不得销售。
版权所有,侵权必究。侵权举报电话:010-62782989　13701121933

图书在版编目(CIP)数据

再也不踩坑的 kubernetes 实战指南 / 杜宽编著. —北京:清华大学出版社,2019
ISBN 978-7-302-53480-8

Ⅰ. ①再… Ⅱ. ①杜… Ⅲ. ①Linux 操作系统—程序设计　Ⅳ. ①TP316.85

中国版本图书馆 CIP 数据核字(2019)第 172252 号

责任编辑:王金柱
封面设计:王　翔
责任校对:闫秀华
责任印制:杨　艳

出版发行:清华大学出版社
网　　址:http://www.tup.com.cn, http://www.wqbook.com
地　　址:北京清华大学学研大厦 A 座　　　邮　编:100084
社 总 机:010-62770175　　　　　　　　　　邮　购:010-62786544
投稿与读者服务:010-62776969, c-service@tup.tsinghua.edu.cn
质 量 反 馈:010-62772015, zhiliang@tup.tsinghua.edu.cn

印 装 者:北京嘉实印刷有限公司
经　　销:全国新华书店
开　　本:190mm×260mm　　印　张:19.5　　字　数:499 千字
版　　次:2019 年 10 月第 1 版　　　　　　　印　次:2019 年 10 月第 1 次印刷
定　　价:89.00 元

产品编号:084076-01

前　言

　　Kubernetes 在近几年，乃至未来 5 到 10 年，都会是技术圈一个很火的名词，Kubernetes 由谷歌（Google）开源，它构建在谷歌 15 年生产环境经验的基础之上，开源的背后有着来自社区的强大技术团队共同维护和更新。Kubernetes 的诞生不仅解决了公司架构带来的问题，而且也大大减少了运维成本，可以轻轻松松管理上万个容器节点。目前很多公司都在致力于对容器和 Kubernetes 的推进，将公司现有的业务拆分为微服务，然后对其进行容器化，所以目前对容器和容器编排工具的学习，是每个技术人员义不容辞的责任。

　　本书主要以 Kubernetes 实战为主，大部分内容都是基于公司实际的应用场景，可直接用于公司生产线上。本书第一章主要讲解 Kubernetes 的高可用安装，分为 kubeadm 和二进制安装方式，可以让读者先入为主，了解 Kubernetes 的架构模式，更快入手。第二章结合实操讲解 Kubernetes 的理论知识。第三章是应用篇，主要讲解的是公司一些常用的应用，并且对其进行容器化，然后部署在 Kubernetes 集群中，实现高可用，使读者对 Kubernetes 的各个组件和资源有一个更好的认识。第 4 章主要讲解持续集成和持续部署，这也是 DevOps 中很重要的一部分，本章主要讲解的是 Jenkins 最新的功能 Pipeline 的使用，从 Pipeline 的语法入手到项目的实操，对传统 Java 和 Spring Cloud 应用的容器化以及自动化构建部署，学习完本章内容读者可以更好地在公司业务上实现以流水线为基础的持续部署。第 5 章主要讲解 Kubernetes 的 Ingress，作为 Kubernetes 集群的入口，使用常见的 Nginx 作为 ingress，会使运维人员更加得心应手，也讲解了 Nginx 的一些常用配置，以适用于生产环境的各种需求和问题。最后一章讲解的是目前很火的 Server Mesh，主要讲解了 Istio 的常用配置，比如使用 Istio 代替微服务架构中的网络功能、实现限速、分流和路由等。本书可以让读者迅速进入到 Kubernetes 的世界，本书的实操内容非常详细，解决了 Kubernetes 在使用和安装过程中遇到的大量问题，能让读者以较短的时间将其应用到公司业务上。

　　本书还配备了完整的命令代码，请扫描下述二维码下载：

由于笔者水平所限，书中难免会出现缪误，请业界高手专家不吝指教，也欢迎各位读者朋友给笔者发邮件进行交流。

邮件地址：dukuan.china@gmail.com

杜宽
2019.5.23

目 录

第1章 Kubernetes 高可用安装 1

1.1 kubeadm 高可用安装 K8S 集群（1.11.x 和 1.12.x）... 1
- 1.1.1 基本环境配置 2
- 1.1.2 内核升级 3
- 1.1.3 基本组件安装 5
- 1.1.4 集群初始化 6
- 1.1.5 Calico 组件的安装 13
- 1.1.6 高可用 Master 14
- 1.1.7 Node 节点的配置 15
- 1.1.8 Metrics-Server 部署 16
- 1.1.9 Dashboard 部署 17

1.2 Kubeadm 高可用安装 K8S 集群（1.13.x 和 1.14.x）... 19
- 1.2.1 基本组件的安装 19
- 1.2.2 集群初始化 20
- 1.2.3 Calico 组件的安装 22
- 1.2.4 高可用 Master 22
- 1.2.5 Node 节点的配置 25

1.3 二进制高可用安装 K8S 集群（1.13.x 和 1.14.x）.. 26
- 1.3.1 基本组件安装 27
- 1.3.2 CNI 安装 28
- 1.3.3 生成证书 29
- 1.3.4 系统组件配置 33
- 1.3.5 TLS Bootstrapping 配置 40
- 1.3.6 Node 节点的配置 42
- 1.3.7 Kube-Proxy 配置 45
- 1.3.8 Calico 配置 47
- 1.3.9 CoreDNS 的配置 48
- 1.3.10 Metrics-Server 配置 49
- 1.3.11 Dashboard 配置 49

1.4 小结 51

第 2 章 Docker 及 Kubernetes 基础 52

2.1 Docker 基础 52
2.1.1 Docker 介绍 52
2.1.2 Docker 基本命令 52
2.1.3 Dockerfile 的编写 59

2.2 Kubernetes 基础 61
2.2.1 Master 节点 61
2.2.2 Node 节点 62
2.2.3 Pod 62
2.2.4 Label 和 Selector 66
2.2.5 Replication Controller 和 ReplicaSet 68
2.2.6 Deployment 70
2.2.7 StatefulSet 77
2.2.8 DaemonSet 87
2.2.9 ConfigMap 91
2.2.10 Secret 99
2.2.11 HPA 105
2.2.12 Storage 107
2.2.13 Service 120
2.2.14 Ingress 124
2.2.15 Taint 和 Toleration 127
2.2.16 RBAC 131
2.2.17 CronJob 138

2.3 小结 140

第 3 章 Kubernetes 常见应用安装 141

3.1 安装 GFS 到 K8S 集群中 141
3.1.1 准备工作 141
3.1.2 创建 GFS 集群 142
3.1.3 创建 Heketi 服务 142
3.1.4 创建 GFS 集群 143
3.1.5 创建 StorageClass 148
3.1.6 测试使用 GFS 动态存储 148
3.1.7 测试数据 149
3.1.8 测试 Deployment 150

3.2 安装 Helm 到 K8S 集群中 153
3.2.1 基本概念 153
3.2.2 安装 Helm 154
3.2.3 Helm 的使用 155

- 3.3 安装 Redis 集群模式到 K8S 集群中 156
  - 3.3.1 各文件介绍 156
  - 3.3.2 创建 Redis 命名空间 161
  - 3.3.3 创建 Redis 集群 PV 161
  - 3.3.4 创建集群 161
  - 3.3.5 创建 slot 162
- 3.4 安装 RabbitMQ 集群到 K8S 集群中 163
  - 3.4.1 各文件解释 163
  - 3.4.2 配置 NFS 167
  - 3.4.3 创建集群 167
  - 3.4.4 查看资源 167
  - 3.4.5 访问测试 168
- 3.5 安装 GitLab 到 K8S 集群中 168
  - 3.5.1 各文件介绍 169
  - 3.5.2 创建 GitLab 173
  - 3.5.3 访问 GitLab 173
  - 3.5.4 创建项目 174
  - 3.5.5 创建用户权限 176
  - 3.5.6 添加 SSH Key 178
  - 3.5.7 项目开发 179
- 3.6 安装 Jenkins 到 K8S 集群中 182
  - 3.6.1 各文件介绍 182
  - 3.6.2 安装 Jenkins 184
  - 3.6.3 访问 Jenkins 185
- 3.7 安装 Harbor 到 K8S 集群中 186
  - 3.7.1 安装 Harbor 186
  - 3.7.2 访问 Harbor 189
  - 3.7.3 在 K8S 中使用 Harbor 190
- 3.8 安装 Prometheus+Grafana 到 K8S 集群中 192
  - 3.8.1 修改配置信息 192
  - 3.8.2 一键安装 Prometheus 192
  - 3.8.3 验证安装 194
  - 3.8.4 访问测试 195
  - 3.8.5 卸载 197
  - 3.8.6 监控 ElasticSearch 集群 197
  - 3.8.7 监控报警配置实战 203
- 3.9 安装 EFK 到 K8S 集群中 205
  - 3.9.1 对节点打标签（Label） 205
  - 3.9.2 创建持久化卷 206

3.9.3 创建集群 206
3.9.4 访问 Kibana 207
3.10 小结 208

第 4 章 持续集成与持续部署 209

4.1 CI/CD 介绍 209
 4.1.1 CI 和 CD 的区别 209
 4.1.2 持续集成（CI） 210
 4.1.3 持续交付（CD） 210
 4.1.4 持续部署 210
4.2 Jenkins 流水线介绍 211
 4.2.1 什么是流水线 211
 4.2.2 Jenkins 流水线概念 211
 4.2.3 声明式流水线 212
 4.2.4 脚本化流水线 212
 4.2.5 流水线示例 213
4.3 Pipeline 语法 214
 4.3.1 声明式流水线 214
 4.3.2 脚本化流水线 223
4.4 Jenkinsfile 的使用 224
 4.4.1 创建 Jenkinsfile 224
 4.4.2 处理 Jenkinsfile 227
4.5 GitLab+ Jenkins +Harbor+Kubernetes 集成应用 233
 4.5.1 基本概念 233
 4.5.2 基本配置 233
 4.5.3 新建任务（Job） 235
 4.5.4 Jenkins 凭据的使用 236
4.6 自动化构建 Java 应用 238
 4.6.1 定义 Dockerfile 238
 4.6.2 定义 Jenkinsfile 238
 4.6.3 定义 Deployment 240
 4.6.4 Harbor 项目创建 241
 4.6.5 创建任务（Job） 242
 4.6.6 执行构建 243
4.7 自动化构建 NodeJS 应用 247
 4.7.1 定义 Dockerfile 247
 4.7.2 定义 Deployment 247
 4.7.3 定义 Jenkinsfile 248
4.8 自动化构建 Spring Cloud 应用 250

	4.8.1 自动化构建 Eureka 250
	4.8.2 自动化构建 Config 255
	4.8.3 自动化构建 Zuul 259

4.9 Webhook 介绍 262
    4.9.1 安装 Webhook 插件 262
    4.9.2 配置 Jenkins 263
    4.9.3 配置 GitLab 263

4.10 自动化构建常见问题的解决 264
    4.10.1 解决代码拉取速度慢的问题 265
    4.10.2 解决 Maven 构建慢的问题 266
    4.10.3 解决 NPM Install 的问题 267

4.11 小结 269

第 5 章 Nginx Ingress 安装与配置 270

5.1 Nginx Ingress 的安装 270
5.2 Nginx Ingress 的简单使用 271
5.3 Nginx Ingress Redirect 272
5.4 Nginx Ingress Rewrite 273
5.5 Nginx Ingress 错误代码重定向 274
5.6 Nginx Ingress SSL 274
5.7 Nginx Ingress 匹配请求头 275
5.8 Nginx Ingress 基本认证 277
5.9 Nginx Ingress 黑/白名单 278
    5.9.1 配置黑名单 278
    5.9.2 配置白名单 278

5.10 Nginx Ingress 速率限制 279
5.11 使用 Nginx 实现灰度/金丝雀发布 280
    5.11.1 创建 v1 版本 280
    5.11.2 创建 v2 版本 281
    5.11.3 创建 Ingress 281
    5.11.4 测试灰度发布 282

5.12 小结 282

第 6 章 Server Mesh 服务网格 283

6.1 服务网格的基本概念 283
6.2 服务网格产品 284
6.3 Istio 介绍 285
    6.3.1 Istio 架构 285
    6.3.2 名词解释 286

6.3.3 流量管理 286
6.4 Istio 的安装 290
    6.4.1 安装文件下载 290
    6.4.2 安装 Istio 290
    6.4.3 配置自动注入 sidecar 291
6.5 Istio 配置请求路由 291
6.6 Istio 熔断 293
    6.6.1 创建测试用例 293
    6.6.2 配置熔断规则 293
    6.6.3 测试熔断 294
6.7 Istio 故障注入 295
    6.7.1 基于 HTTP 延迟触发故障 295
    6.7.2 使用 HTTP Abort 触发故障 299
6.8 Istio 速率限制 299
    6.8.1 配置速率限制 299
    6.8.2 测试速率限制 302
6.9 小结 302

第 1 章

Kubernetes 高可用安装

Kubernetes 作为容器编排的佼佼者，已经被很多公司认可，使用越来越广泛。Kubernetes 基于允许谷歌（Google）每周运行数十亿个容器的原则而设计，无论应用运行在本地还是全球任何地域，Kubernetes 的灵活性都可以随着需求复杂度的不断增加，持续、轻松地对外提供服务。

本章首先介绍 Kubernetes 的安装，之所以从安装入手，是因为在生产线上，首要任务是要有一套自己的集群，才能更好地开展后期的工作，同时在部署的过程中，也会接触到 Kubernetes 各个组件，可以达到先入为主的效果。此外，在已有集群的情况下，再去了解相关基础，印象就会更加深刻。当然，读者也可以从第 2 章基础知识开始，根据自己的喜好或者需求进行选择。

本章会讲到两种安装方式：kubeadm 和二进制安装方式，kubeadm 安装较为简单，非常适合新手学习和熟悉 Kubernetes；二进制安装方式较为复杂，但是它是到目前为止推荐在生产环境中使用的安装方式，虽然 kubeadm 已经成为 Kubernetes 官方默认的安装方式，但是仍然不建议在生产环境中使用。经测试，当集群全部宕机（发生此种情况的机会很小）时，二进制安装方式恢复能力较强，速度较快。

Kubeadm 安装 1.11 版本和 1.12 版本类似，1.13 版本改动较大，但 1.13 和 1.14 版本安装类似，且 1.13 和 1.14 版本的安装过程更为简单。以下介绍 Kubeadm 安装 1.11、1.12、1.13 和 1.14 版，通过二进制方式安装 1.13 和 1.14 版。笔者写此书时官方最新稳定版本为 1.14 版，在实际安装时，可以选择其中一个版本进行安装即可，本章所用代码在对应的 chap01 目录下。

1.1 kubeadm 高可用安装 K8S 集群（1.11.x 和 1.12.x）

本节主要演示使用 Kubeadm 安装 Kubernetes 高可用集群，笔者公司大部分线下测试环境均采用 Kubeadm 安装，这也是目前官方默认的安装方式，比二进制安装方式更加简单，可以让初学者快速上手并测试。目前 GitHub 上也有很多基于 Ansible 的自动化安装方式，但是为了更好地学习 Kubernetes，还是建议体验一下 Kubernetes 的手动安装过程，以熟悉 Kubernetes 的各个组件。

截止到本书截稿前，官方最新的稳定版本为 1.14，本章的内容会涉及到 1.11、1.12、1.13 和 1.14 版本的安装，对于 Kubeadm 来说，安装 1.11.x 版本和 1.12.x 类似，只需更改对应的 Kubernetes 版本号即可。本节主要演示的是 1.11.x 和 1.12.x 的安装。

1.1.1 基本环境配置

本次安装使用 5 台 Linux 服务器，系统版本为 CentOS 7.5，分为 3 台 Master、2 台 Node，其中 Node 的配置相同。Master 节点主要部署的组件有 KeepAlived、HAProxy、Etcd、Kubelet、APIServer、Controller、Scheduler，Node 节点主要部署的为 Kubelet，详情见表 1-1。其中的概念可以参考第 2 章 Docker 和 Kubernetes 基础部分的内容。

表 1-1 高可用 Kubernetes 集群规划

主机名	IP 地址	说明
K8S-master01 ~ 03	192.168.20.20 ~ 22	master 节点×3
K8S-master-lb	192.168.20.10	keepalived 虚拟 IP
K8S-node01 ~ 02	192.168.20.30 ~ 31	worker 节点×2

各节点通信采用主机名的方式，这种方式与 IP 地址相比较更具有扩展性。以下介绍具体的安装步骤。

所有节点配置 hosts，修改 /etc/hosts 如下：

```
[root@K8S-master01 ~]# cat /etc/hosts
192.168.20.20 K8S-master01
192.168.20.21 K8S-master02
192.168.20.22 K8S-master03
192.168.20.10 K8S-master-lb
192.168.20.30 K8S-node01
192.168.20.31 K8S-node02
```

所有节点关闭防火墙、selinux、dnsmasq、swap（如果开启防火墙需要开放对应的端口，配置较为复杂）。如果在云上部署，可以通过安全组进行安全配置。服务器配置如下：

```
systemctl disable --now firewalld
systemctl disable --now dnsmasq
systemctl disable --now NetworkManager
```

关闭 Selinux

```
setenforce 0
```

将 /etc/sysconfig/selinux 文件中的 SELINUX 改为 disabled：

```
[root@k8s-node01 ~]# grep -vE "#|^$" /etc/sysconfig/selinux
SELINUX=disabled
SELINUXTYPE=targeted
```

关闭 swap 分区

```
swapoff -a && sysctl -w vm.swappiness=0
```

注释 swap 挂载选项：

```
[root@k8s-node01 ~]# grep "swap" /etc/fstab
#UUID=afb91fab-ca66-4bb6-b5ce-a9366abf18e3 swap                swap
defaults        0 0
```

所有节点同步时间。所有节点同步时间是必须的，并且需要加到开机自启动和计划任务中，如果节点时间不同步，会造成 Etcd 存储 Kubernetes 信息的键-值（key-value）数据库同步数据不正常，也会造成证书出现问题。时间同步配置如下：

```
ln -sf /usr/share/zoneinfo/Asia/Shanghai /etc/localtime
echo 'Asia/Shanghai' >/etc/timezone
ntpdate time2.aliyun.com
# 加入到 crontab
*/5 * * * * ntpdate time2.aliyun.com
# 加入到开机自动同步,/etc/rc.local
ntpdate time2.aliyun.com
```

所有节点配置 limit：

```
ulimit -SHn 65535
```

Master01 节点免密钥登录其他节点，安装过程中生成配置文件和证书均在 Master01 上操作，集群管理也在 Master01 上操作，阿里云或者 AWS 上需要单独一台 kubectl 服务器。密钥配置如下：

```
ssh-keygen -t rsa
for i in K8S-master01 K8S-master02 K8S-master03 K8S-node01 K8S-node02;do ssh-copy-id -i .ssh/id_rsa.pub $i;done
```

Master01 节点下载安装文件，本节所用的安装文件均放在 chap01/1.1 目录中。

在源码中的 repo 目录配置使用的是国内仓库源，将其复制到所有节点：

```
for i in K8S-master01 K8S-master02 K8S-master03 K8S-node01 K8S-node02;do scp -r chap01/1.1/repo/ $i:/opt ;done
```

所有节点配置 repo 源：

```
cd /etc/yum.repos.d
mkdir bak
mv *.repo bak/
mv /opt/repo/* .
```

所有节点升级系统并重启，此处升级没有升级内核，下节会单独升级内核：

```
yum install wget jq psmisc vim net-tools -y
yum update -y --exclude=kernel* && reboot
```

1.1.2 内核升级

在安装过程中，很多文档及网上资源不会提及到内核升级的部分，但升级内核可以减少一些不必要的 Bug，也是安装过程中颇为重要的一步。

本例升级的内核版本为 4.18，采用 rpm 的安装方式，Master01 节点下载内核升级包：

```
wget
```

```
http://mirror.rc.usf.edu/compute_lock/elrepo/kernel/el7/x86_64/RPMS/kernel-ml-4.18.9-1.el7.elrepo.x86_64.rpm
    wget
http://mirror.rc.usf.edu/compute_lock/elrepo/kernel/el7/x86_64/RPMS/kernel-ml-devel-4.18.9-1.el7.elrepo.x86_64.rpm
```

将内核升级包复制到其他节点：

```
    for i in K8S-master02 K8S-master03 K8S-node01 K8S-node02;do scp kernel-ml-4.18.9-1.el7.elrepo.x86_64.rpm kernel-ml-devel-4.18.9-1.el7.elrepo.x86_64.rpm $i:/root/ ; done
```

所有节点升级内核：

```
    yum localinstall -y kernel-ml*
```

所有节点修改内核启动顺序：

```
    grub2-set-default  0 && grub2-mkconfig -o /etc/grub2.cfg
    grubby --args="user_namespace.enable=1" --update-kernel="$(grubby --default-kernel)"
```

所有节点重启：

```
    reboot
```

所有节点再次启动后确认内核版本：

```
    uname -r
```

本书的 Kube-Proxy 均采用 ipvs 模式，该模式也是新版默认支持的代理模式，性能比 iptables 要高，如果服务器未配置安装 ipvs，将转换为 iptables 模式。所有节点安装 ipvsadm：

```
    yum install ipvsadm ipset sysstat conntrack libseccomp -y
```

所有节点配置 ipvs 模块，在内核 4.19 版本 nf_conntrack_ipv4 已经改为 nf_conntrack，本例安装的内核为 4.18，使用 nf_conntrack_ipv4 即可：

```
    modprobe -- ip_vs
    modprobe -- ip_vs_rr
    modprobe -- ip_vs_wrr
    modprobe -- ip_vs_sh
    modprobe -- nf_conntrack_ipv4
    modprobe -- ip_tables
    modprobe -- ip_set
    modprobe -- xt_set
    modprobe -- ipt_set
    modprobe -- ipt_rpfilter
    modprobe -- ipt_REJECT
    modprobe -- ipip
```

检查是否加载，并将其加入至开机自动加载（在目录 /etc/sysconfig/modules/ 下创建一个 k8s.modules 写上上述命令即可）：

```
    [root@K8S-master01 ~]# lsmod | grep -e ip_vs -e nf_conntrack_ipv4
    nf_conntrack_ipv4      16384  23
    nf_defrag_ipv4         16384  1 nf_conntrack_ipv4
    nf_conntrack          135168  10
```

```
xt_conntrack,nf_conntrack_ipv6,nf_conntrack_ipv4,nf_nat,nf_nat_ipv6,ipt_MASQUE
RADE,nf_nat_ipv4,xt_nat,nf_conntrack_netlink,ip_vs
```

开启一些 K8S 集群中必须的内核参数,所有节点配置 K8S 内核:

```
cat <<EOF > /etc/sysctl.d/K8S.conf
net.ipv4.ip_forward = 1
net.bridge.bridge-nf-call-iptables = 1
fs.may_detach_mounts = 1
vm.overcommit_memory=1
vm.panic_on_oom=0
fs.inotify.max_user_watches=89100
fs.file-max=52706963
fs.nr_open=52706963
net.netfilter.nf_conntrack_max=2310720

net.ipv4.tcp_keepalive_time = 600
net.ipv4.tcp_keepalive_probes = 3
net.ipv4.tcp_keepalive_intvl =15
net.ipv4.tcp_max_tw_buckets = 36000
net.ipv4.tcp_tw_reuse = 1
net.ipv4.tcp_max_orphans = 327680
net.ipv4.tcp_orphan_retries = 3
net.ipv4.tcp_syncookies = 1
net.ipv4.tcp_max_syn_backlog = 16384
net.ipv4.ip_conntrack_max = 65536
net.ipv4.tcp_max_syn_backlog = 16384
net.ipv4.tcp_timestamps = 0
net.core.somaxconn = 16384
EOF
sysctl --system
```

1.1.3 基本组件安装

本节主要安装的是集群中用到的各种组件,比如 Docker-ce、Kubernetes 各组件等。
查看可用 docker-ce 版本:

```
yum list docker-ce.x86_64 --showduplicates | sort -r
```

目前官方经过测试的 Docker 版本有 1.11.1、1.12.1、1.13.1、17.03、17.06、17.09、18.06,可自行选择。

这里安装的 Docker 版本为 17.09,其他版本自行更改即可,所有节点安装 docker-ce-17.09:

```
yum -y install docker-ce-17.09.1.ce-1.el7.centos
```

和 docker-ce 一样,首先查看可用 Kubeadm 组件版本:

```
yum list kubeadm.x86_64 --showduplicates | sort -r
```

所有节点安装 K8S 组件。本例安装的为 1.12.3,可以将版本改为 1.11.x 或 1.12.x,请自行选择:

```
yum install -y kubeadm-1.12.3-0.x86_64 kubectl-1.12.3-0.x86_64 kubelet-1.12.3-0.x86_64
```

所有节点设置开机自启动 Docker：

```
systemctl enable --now docker
```

默认配置的 pause 镜像使用 gcr.io 仓库，国内可能无法访问，所以这里配置 Kubelet 使用阿里云的 pause 镜像，使用 kubeadm 初始化时会读取该文件的变量：

```
DOCKER_CGROUPS=$(docker info | grep 'Cgroup' | cut -d' ' -f3)
cat >/etc/sysconfig/kubelet<<EOF
KUBELET_EXTRA_ARGS="--cgroup-driver=$DOCKER_CGROUPS
--pod-infra-container-image=registry.cn-hangzhou.aliyuncs.com/google_containers/pause-amd64:3.1"
    EOF
```

设置 Kubelet 开机自启动：

```
systemctl daemon-reload
systemctl enable --now kubelet
```

1.1.4　集群初始化

本节进行 Kubernetes 集群初始化，主要目的是生成集群中用到的证书和配置文件。在二进制安装过程中，证书和配置文件需要自行生成。

本例高可用采用的是 HAProxy+Keepalived，HAProxy 和 KeepAlived 以守护进程的方式在所有 Master 节点部署。通过 yum 安装 HAProxy 和 KeepAlived：

```
yum install keepalived haproxy -y
```

所有 Master 节点配置 HAProxy（详细配置参考 HAProxy 文档，所有 Master 节点的 HAProxy 配置相同）：

```
[root@K8S-master01 etc]# mkdir /etc/haproxy
[root@K8S-master01 etc]# cat /etc/haproxy/haproxy.cfg
global
  maxconn  2000
  ulimit-n 16384
  log  127.0.0.1 local0 err
  stats timeout 30s

defaults
  log global
  mode  http
  option  httplog
  timeout connect 5000
  timeout client  50000
  timeout server  50000
  timeout http-request 15s
  timeout http-keep-alive 15s

frontend monitor-in
  bind *:33305
  mode http
  option httplog
```

```
    monitor-uri /monitor

  listen stats
    bind     *:8006
    mode     http
    stats    enable
    stats    hide-version
    stats    uri         /stats
    stats    refresh     30s
    stats    realm       Haproxy\ Statistics
    stats    auth        admin:admin

  frontend K8S-master
    bind 0.0.0.0:16443
    bind 127.0.0.1:16443
    mode tcp
    option tcplog
    tcp-request inspect-delay 5s
    default_backend K8S-master

  backend K8S-master
    mode tcp
    option tcplog
    option tcp-check
    balance roundrobin
    default-server inter 10s downinter 5s rise 2 fall 2 slowstart 60s maxconn 250 maxqueue 256 weight 100
      server K8S-master01    192.168.20.20:6443  check
      server K8S-master02    192.168.20.21:6443  check
      server K8S-master03    192.168.20.22:6443  check
```

所有 Master 节点配置 KeepAlived。注意修改 interface（服务器网卡）、priority（优先级，不同即可）、mcast_src_ip（本机 IP），详细配置参考 keepalived 文档。

Master01 节点的配置：

```
[root@K8S-master01 etc]# mkdir /etc/keepalived

[root@K8S-master01 ~]# cat /etc/keepalived/keepalived.conf
! Configuration File for keepalived
global_defs {
    router_id LVS_DEVEL
}
vrrp_script chk_apiserver {
    script "/etc/keepalived/check_apiserver.sh"
    interval 2
    weight -5
    fall 3
    rise 2
}
vrrp_instance VI_1 {
    state MASTER
    interface ens160
    mcast_src_ip 192.168.20.20
    virtual_router_id 51
    priority 100
```

```
        advert_int 2
        authentication {
            auth_type PASS
            auth_pass K8SHA_KA_AUTH
        }
        virtual_ipaddress {
            192.168.20.10
        }
#       track_script {
#           chk_apiserver
#       }
}
```

Master02 节点的配置：

```
! Configuration File for keepalived
global_defs {
    router_id LVS_DEVEL
}
vrrp_script chk_apiserver {
    script "/etc/keepalived/check_apiserver.sh"
    interval 2
    weight -5
    fall 3
    rise 2
}
vrrp_instance VI_1 {
    state BACKUP
    interface ens160
    mcast_src_ip 192.168.20.21
    virtual_router_id 51
    priority 101
    advert_int 2
    authentication {
        auth_type PASS
        auth_pass K8SHA_KA_AUTH
    }
    virtual_ipaddress {
        192.168.20.10
    }
#   track_script {
#       chk_apiserver
#   }
}
```

Master03 节点的配置：

```
! Configuration File for keepalived
global_defs {
    router_id LVS_DEVEL
}
vrrp_script chk_apiserver {
    script "/etc/keepalived/check_apiserver.sh"
    interval 2
    weight -5
    fall 3
```

```
      rise 2
}
vrrp_instance VI_1 {
    state BACKUP
    interface ens160
    mcast_src_ip 192.168.20.22
    virtual_router_id 51
    priority 102
    advert_int 2
    authentication {
        auth_type PASS
        auth_pass K8SHA_KA_AUTH
    }
    virtual_ipaddress {
        192.168.20.10
    }
#    track_script {
#        chk_apiserver
#    }
}
```

注意，上述的健康检查是关闭的，集群建立完成后再开启：

```
#    track_script {
#        chk_apiserver
#    }
```

配置 KeepAlived 健康检查文件：

```
[root@K8S-master01 keepalived]# cat /etc/keepalived/check_apiserver.sh
#!/bin/bash

function check_apiserver() {
  for ((i=0;i<5;i++));do
    apiserver_job_id=$(pgrep kube-apiserver)
    if [[ ! -z $apiserver_job_id ]];then
      return
    else
      sleep 2
    fi
    apiserver_job_id=0
  done
}

# 1: running 0: stopped
check_apiserver
if [[ $apiserver_job_id -eq 0 ]]; then
    /usr/bin/systemctl stop keepalived
    exit 1
else
    exit 0
fi
```

启动 haproxy 和 keepalived

```
[root@K8S-master01 keepalived]# systemctl enable --now haproxy
```

```
[root@K8S-master01 keepalived]# systemctl enable --now keepalived
```

> **注　意**
>
> 高可用方式不一定非要采用 HAProxy 和 KeepAlived，在云上的话可以使用云上的负载均衡，比如在阿里云上可以使用阿里云内部的 SLB，就无须再配置 HAProxy 和 KeepAlived，只需要将对应的 VIP 改成 SLB 的地址即可。在企业内部可以使用 F5 硬件负载均衡，反向代理到每台 Master 节点的 6443 端口即可。

Kubeadm 的安装方式可以配合使用 kubeadm-config 文件来初始化集群，所以需要提前创建各 Master 节点的 kubeadm-config。由于国内网络的问题，需要将集群镜像的仓库地址改成 imageRepository: registry.cn-hangzhou.aliyuncs.com/google_containers。

各 Master 节点的配置文件如下：

Master01：

```yaml
apiVersion: kubeadm.K8S.io/v1alpha2
kind: MasterConfiguration
kubernetesVersion: v1.12.3
imageRepository: registry.cn-hangzhou.aliyuncs.com/google_containers
api:
  advertiseAddress: 192.168.20.20
  controlPlaneEndpoint: K8S-master-lb:16443
controllerManagerExtraArgs:
  node-monitor-grace-period: 10s
  pod-eviction-timeout: 10s

apiServerCertSANs:
- 192.168.20.20
- 192.168.20.21
- 192.168.20.22
- K8S-master-lb
- K8S-master01
- K8S-master02
- K8S-master03
- 192.168.20.10
etcd:
  local:
    extraArgs:
      listen-client-urls: "https://127.0.0.1:2379,https://192.168.20.20:2379"
      advertise-client-urls: "https://192.168.20.20:2379"
      listen-peer-urls: "https://192.168.20.20:2380"
      initial-advertise-peer-urls: "https://192.168.20.20:2380"
      initial-cluster: "K8S-master01=https://192.168.20.20:2380"
    serverCertSANs:
      - K8S-master01
      - 192.168.20.20
    peerCertSANs:
      - K8S-master01
      - 192.168.20.20
networking:
  podSubnet: "172.168.0.0/16"
```

```
#kubeProxy:
#  config:
#    featureGates:
#      SupportIPVSProxyMode: true
#    mode: ipvs
```

Master02：

```
apiVersion: kubeadm.K8S.io/v1alpha2
kind: MasterConfiguration
kubernetesVersion: v1.12.3

imageRepository: registry.cn-hangzhou.aliyuncs.com/google_containers
api:
  advertiseAddress: 192.168.20.21
  controlPlaneEndpoint: 192.168.20.10:16443
controllerManagerExtraArgs:
  node-monitor-grace-period: 10s
  pod-eviction-timeout: 10s
apiServerCertSANs:
- K8S-master02
- K8S-master02
- K8S-master03
- K8S-master-lb
- 192.168.20.20
- 192.168.20.21
- 192.168.20.22
- 192.168.20.10
etcd:
  local:
    extraArgs:
      listen-client-urls: "https://127.0.0.1:2379,https://192.168.20.21:2379"
      advertise-client-urls: "https://192.168.20.21:2379"
      listen-peer-urls: "https://192.168.20.21:2380"
      initial-advertise-peer-urls: "https://192.168.20.21:2380"
      initial-cluster: "K8S-master01=https://192.168.20.20:2380,K8S-master02=https://192.168.20.21:2380"
      initial-cluster-state: existing
    serverCertSANs:
      - K8S-master02
      - 192.168.20.21
    peerCertSANs:
      - K8S-master02
      - 192.168.20.21
networking:
  podSubnet: "172.168.0.0/16"

#kubeProxy:
#  config:
#    featureGates:
#      SupportIPVSProxyMode: true
#    mode: ipvs
```

Master03：

```yaml
apiVersion: kubeadm.K8S.io/v1alpha2
kind: MasterConfiguration
kubernetesVersion: v1.12.3

imageRepository: registry.cn-hangzhou.aliyuncs.com/google_containers
api:
  advertiseAddress: 192.168.20.22
  controlPlaneEndpoint: 192.168.20.10:16443
controllerManagerExtraArgs:
  node-monitor-grace-period: 10s
  pod-eviction-timeout: 10s
apiServerCertSANs:
- K8S-master02
- K8S-master02
- K8S-master03
- K8S-master-lb
- 192.168.20.20
- 192.168.20.21
- 192.168.20.22
- 192.168.20.10
etcd:
  local:
    extraArgs:
      listen-client-urls: "https://127.0.0.1:2379,https://192.168.20.22:2379"
      advertise-client-urls: "https://192.168.20.22:2379"
      listen-peer-urls: "https://192.168.20.22:2380"
      initial-advertise-peer-urls: "https://192.168.20.22:2380"
      initial-cluster: "K8S-master01=https://192.168.20.20:2380,K8S-master02=https://192.168.20.21:2380,K8S-master03=https://192.168.20.22:2380"
      initial-cluster-state: existing
    serverCertSANs:
      - K8S-master03
      - 192.168.20.22
    peerCertSANs:
      - K8S-master03
      - 192.168.20.22
networking:
  # This CIDR is a calico default. Substitute or remove for your CNI provider.
  podSubnet: "172.168.0.0/16"
#kubeProxy:
#  config:
#    featureGates:
#      SupportIPVSProxyMode: true
#    mode: ipvs
```

所有 Master 节点提前下载镜像，可以节省初始化时间：

```
kubeadm config images pull --config /root/kubeadm-config.yaml
```

Master01 节点初始化，初始化以后会在/etc/kubernetes 目录下生成对应的证书和配置文件，之后其他 Master 节点加入 Master01 即可：

```
kubeadm init --config /root/kubeadm-config.yaml
```

如果初始化失败，重置后再次初始化，命令如下：

```
kubeadm reset
```

初始化成功以后，会产生 Token 值，用于其他节点加入时使用，因此要记录下初始化成功生成的 token 值（令牌值）：

```
kubeadm join 192.168.20.10:16443 --token sj5ymu.8nl0m093pu0zgb37
--discovery-token-ca-cert-hash
sha256:ec16a73977f6c5a99f4556c17de29d8ac3b57288bb48dc9c491bd23744d011bd
```

所有 Master 节点配置环境变量，用于访问 Kubernetes 集群：

```
cat <<EOF >> /root/.bashrc
export KUBECONFIG=/etc/kubernetes/admin.conf
EOF
source /root/.bashrc
```

查看节点状态：

```
[root@K8S-master01 ~]# kubectl get nodes
NAME             STATUS     ROLES    AGE    VERSION
K8S-master01     NotReady   master   14m    v1.12.3
```

采用初始化安装方式，所有的系统组件均以容器的方式运行并且在 kube-system 命名空间内，此时可以查看 Pod 状态：

```
[root@K8S-master01 ~]# kubectl get pods -n kube-system -o wide
  NAME                                         READY   STATUS    RESTARTS   AGE     IP             NODE
  coredns-777d78ff6f-kstsz                     0/1     Pending   0          14m     <none>         <none>
  coredns-777d78ff6f-rlfr5                     0/1     Pending   0          14m     <none>         <none>
  etcd-K8S-master01                            1/1     Running   0          14m     192.168.20.20  K8S-master01
  kube-apiserver-K8S-master01                  1/1     Running   0          13m     192.168.20.20  K8S-master01
  kube-controller-manager-K8S-master01         1/1     Running   0          13m     192.168.20.20  K8S-master01
  kube-proxy-8d4qc                             1/1     Running   0          14m     192.168.20.20  K8S-master01
  kube-scheduler-K8S-master01                  1/1     Running   0          13m     192.168.20.20  K8S-master01
```

1.1.5 Calico 组件的安装

Calico 作为 Kubernetes 集群的网络组件，主要用来为 Kubernetes 创建和管理一个三层网络，为每个容器分配一个可路由的 IP 地址，实现集群中 Pod 之间的通信。

安装 Calico 3.3.2：

```
kubectl create -f calico/calico.yaml -f calico/upgrade/rbac-kdd.yaml
```

再次查看 Pod 和 Node 节点的状态，可以发现 CoreDNS 已经处于 Running 状态，且 Node 的 STATUS 变成了 Ready 状态，此时表示 Calico 安装成功：

```
kubectl get po,node -n kube-system
```

1.1.6　高可用 Master

本节介绍 Kubernetes 的高可用配置，如果暂时不需要高可用集群可以略过此节，然后将其 VIP 改为 Master01 节点的 IP 地址即可。在生产线上 Master 组件的高可用是很重要的一部分，可用于防止 Master 节点宕机后对集群造成的影响。

复制证书到其他 Master 节点：

```
USER=root
CONTROL_PLANE_IPS="K8S-master02 K8S-master03"
for host in $CONTROL_PLANE_IPS; do
    ssh "${USER}"@$host "mkdir -p /etc/kubernetes/pki/etcd"
    scp /etc/kubernetes/pki/ca.crt "${USER}"@$host:/etc/kubernetes/pki/ca.crt
    scp /etc/kubernetes/pki/ca.key "${USER}"@$host:/etc/kubernetes/pki/ca.key
    scp /etc/kubernetes/pki/sa.key "${USER}"@$host:/etc/kubernetes/pki/sa.key
    scp /etc/kubernetes/pki/sa.pub "${USER}"@$host:/etc/kubernetes/pki/sa.pub
    scp /etc/kubernetes/pki/front-proxy-ca.crt "${USER}"@$host:/etc/kubernetes/pki/front-proxy-ca.crt
    scp /etc/kubernetes/pki/front-proxy-ca.key "${USER}"@$host:/etc/kubernetes/pki/front-proxy-ca.key
    scp /etc/kubernetes/pki/etcd/ca.crt "${USER}"@$host:/etc/kubernetes/pki/etcd/ca.crt
    scp /etc/kubernetes/pki/etcd/ca.key "${USER}"@$host:/etc/kubernetes/pki/etcd/ca.key
    scp /etc/kubernetes/admin.conf "${USER}"@$host:/etc/kubernetes/admin.conf
done
```

和 Master01 一样，在 Master02 节点上提前下载镜像：

```
kubeadm config images pull --config /root/kubeadm-config.yaml
```

在 Master02 节点上创建证书及 kubelet 配置文件：

```
kubeadm alpha phase certs all --config /root/kubeadm-config.yaml
kubeadm alpha phase kubeconfig controller-manager --config /root/kubeadm-config.yaml
kubeadm alpha phase kubeconfig scheduler --config /root/kubeadm-config.yaml
kubeadm alpha phase kubelet config write-to-disk --config /root/kubeadm-config.yaml
kubeadm alpha phase kubelet write-env-file --config /root/kubeadm-config.yaml
kubeadm alpha phase kubeconfig kubelet --config /root/kubeadm-config.yaml
```

重启 Kubelet：

```
systemctl restart kubelet
```

将 Master02 的 Etcd 加入到 Master01 的 Etcd 集群中：

```
    kubectl exec -n kube-system etcd-K8S-master01 -- etcdctl --ca-file
/etc/kubernetes/pki/etcd/ca.crt --cert-file /etc/kubernetes/pki/etcd/peer.crt
--key-file /etc/kubernetes/pki/etcd/peer.key
--endpoints=https://192.168.20.20:2379 member add K8S-master02
https://192.168.20.21:2380
    kubeadm alpha phase etcd local --config /root/kubeadm-config.yaml
```

启动 Master02：

```
kubeadm alpha phase kubeconfig all --config /root/kubeadm-config.yaml
kubeadm alpha phase controlplane all --config /root/kubeadm-config.yaml
kubeadm alpha phase mark-master --config /root/kubeadm-config.yaml
```

配置 Master03 和配置 Master02 的步骤基本一致，除了配置 Etcd 集群时的 IP 地址和主机名信息不一致，其余步骤完全一致。

所有 Master 配置 KUBECONFIG，用于访问集群：

```
mkdir -p $HOME/.kube
cp -i /etc/kubernetes/admin.conf $HOME/.kube/config
chown $(id -u):$(id -g) $HOME/.kube/config
```

1.1.7　Node 节点的配置

Node 节点上主要部署公司的一些业务应用，生产环境中不建议 Master 节点部署系统组件之外的其他 Pod，测试环境可以允许 Master 节点部署 Pod 以节省系统资源。

清理 Node 节点 Kubelet 配置：

```
kubeadm reset
```

使用 kubeadm join 将 Node 节点加入集群，使用的是刚才初始化 Master 生成的 Token：

```
    kubeadm join 192.168.20.10:16443 --token sj5ymu.8n10m093pu0zgb37
--discovery-token-ca-cert-hash
sha256:ec16a73977f6c5a99f4556c17de29d8ac3b57288bb48dc9c491bd23744d011bd
    ......
    [kubelet-start] Activating the kubelet service
    [tlsbootstrap] Waiting for the kubelet to perform the TLS Bootstrap...
    [patchnode] Uploading the CRI Socket information "/var/run/dockershim.sock"
to the Node API object "K8S-node02" as an annotation

    This node has joined the cluster:
    * Certificate signing request was sent to apiserver and a response was received.
    * The Kubelet was informed of the new secure connection details.

    Run 'kubectl get nodes' on the master to see this node join the cluster.
```

所有 Node 节点配置相同，加入后查看节点：

```
kubectl get nodes
[root@K8S-master01 K8S-ha-install]# kubectl get no
NAME            STATUS   ROLES    AGE   VERSION
K8S-master01    Ready    master   13h   v1.12.3
K8S-master02    Ready    master   13h   v1.12.3
K8S-master03    Ready    master   13h   v1.12.3
```

```
K8S-node01       Ready    <none>    11m    v1.12.3
K8S-node02       Ready    <none>    11m    v1.12.3
```

允许 Master 节点部署 Pod，但并不是必需的，生产环境中请勿允许 Master 节点部署系统组件之外的其他 Pod，以免升级集群或维护时对业务造成影响。

```
kubectl taint nodes --all node-role.kubernetes.io/master-
```

1.1.8 Metrics-Server 部署

在新版的 Kubernetes 中系统资源的采集均使用 Metrics-server，可以通过 Metrics 采集节点和 Pod 的内存、磁盘、CPU 和网络的使用率。

所有 Master 节点允许 HPA 通过接口采集数据（新版本默认开启），修改后 Pod 会自动重启：

```
vi /etc/kubernetes/manifests/kube-controller-manager.yaml
- --horizontal-pod-autoscaler-use-rest-clients=false
```

安装 Metrics-server：

```
kubectl apply -f metrics-server/
```

等待几分钟可以查看获取的数据：

```
NAME              CPU(cores)    CPU%    MEMORY(bytes)    MEMORY%
K8S-master01      176m          4%      1673Mi           43%
K8S-master02      180m          4%      1422Mi           37%
K8S-master03      175m          4%      1412Mi           36%
K8S-node01        228m          5%      1299Mi           16%
K8S-node02        424m          10%     1463Mi           18%
You have new mail in /var/spool/mail/root
[root@K8S-master01 K8S-ha-install]# kubectl top pods -n kube-system
NAME                                          CPU(cores)    MEMORY(bytes)
calico-node-2hvmw                             13m           99Mi
calico-node-lbxlm                             38m           99Mi
calico-node-sc92z                             14m           98Mi
calico-node-scqwn                             15m           99Mi
calico-node-sqsm9                             43m           102Mi
coredns-6c66ffc55b-4n6tj                      2m            14Mi
coredns-6c66ffc55b-w6sgh                      2m            14Mi
etcd-K8S-master01                             30m           154Mi
etcd-K8S-master02                             30m           144Mi
etcd-K8S-master03                             23m           144Mi
kube-apiserver-K8S-master01                   34m           752Mi
kube-apiserver-K8S-master02                   24m           622Mi
kube-apiserver-K8S-master03                   22m           612Mi
kube-controller-manager-K8S-master01          24m           88Mi
kube-controller-manager-K8S-master02          1m            14Mi
kube-controller-manager-K8S-master03          1m            14Mi
kube-proxy-cmq84                              8m            27Mi
kube-proxy-f4bkj                              5m            26Mi
kube-proxy-hw47r                              4m            26Mi
kube-proxy-jtd2f                              4m            24Mi
kube-proxy-lssvx                              6m            26Mi
kube-scheduler-K8S-master01                   7m            16Mi
```

```
kube-scheduler-K8S-master02                     10m           15Mi
kube-scheduler-K8S-master03                     9m            20Mi
kubernetes-dashboard-7c7c5ff466-b6957           2m            19Mi
metrics-server-744f6c5f6d-9gb7k                 5m            23Mi
```

1.1.9　Dashboard 部署

Dashboard 用于展示集群中的各类资源，同时也可以通过 Dashboard 实时查看 Pod 的日志和在容器中执行一些命令等。

安装 Dashboard：

```
kubectl apply -f dashboard/
```

安装 heapster，虽然新版的 Kubernetes 用 metrics-server 顶替了 heapster 获取集群资源监控数据，但是 1.x 版本的 Dashboard 还是使用 heapster 获取集群的资源数据，所以部署 Dashboard 1.x 版本还是需要先安装 heapster。如果不需要 Dashboard 展示资源监控数据或者使用 Dashboard 2.x 版本，则可以不安装 heapster。

```
kubectl apply -f heapster/
```

在谷歌浏览器（Chrome）启动文件中加入启动参数，用于解决无法访问 Dashboard 的问题，参考图 1-1。

```
--test-type --ignore-certificate-errors
```

图 1-1　谷歌浏览器 Chrome 的配置

访问 Dashboard：https://192.168.20.10:30000，选择登录方式为令牌（即 token 方式），参考图 1-2。

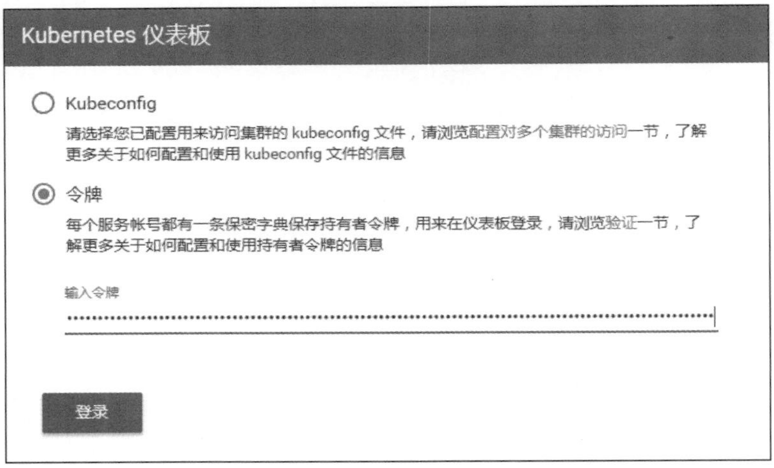

图 1-2　Dashboard 登录方式

查看 token 值：

```
[root@K8S-master01 1.1.1]# kubectl -n kube-system describe secret $(kubectl -n kube-system get secret | grep admin-user | awk '{print $1}')
Name:         admin-user-token-r4vcp
Namespace:    kube-system
Labels:       <none>
Annotations:  kubernetes.io/service-account.name: admin-user
              kubernetes.io/service-account.uid: 2112796c-1c9e-11e9-91ab-000c298bf023

Type:  kubernetes.io/service-account-token

Data
====
ca.crt:     1025 bytes
namespace:  11 bytes
token:      eyJhbGciOiJSUzI1NiIsImtpZCI6IiJ9.eyJpc3MiOiJrdWJlcm5ldGVzL3NlcnZpY2VhY2NvdW50Iiwia3ViZXJuZXRlcy5pby9zZXJ2aWNlYWNjb3VudC9uYW1lc3BhY2UiOiJrdWJlLXN5c3RlbSIsImt1YmVybmV0ZXMuaW8vc2Vydmljb2FjY291bnQvc2VjcmV0Lm5hbWUiOiJhZG1pbi11c2VyLXRva2VuLXI0dmNwIiwia3ViZXJuZXRlcy5pby9zZXJ2aWNlYWNjb3VudC9zZXJ2aWNlLWFjY291bnQubmFtZSI6ImFkbWluLXVzZXIiLCJrdWJlcm5ldGVzLmlvL3NlcnZpY2VhY2NvdW50L3NlcnZpY2UtYWNjb3VudC51aWQiOiIyMTEyNzk2Yy0xYzllLTExZTktOTFhYi0wMDBjMjk4YmYwMjMiLCJzdWIiOiJzeXN0ZW06c2VydmljZWFjY291bnQ6a3ViZS1zeXN0ZW06YWRtaW4tdXNlciJ9.bWYmwgRb-90ydQmyjkbjJjFt8CdO8u6zxVZh-19rdlL_T-n35nKyQIN7hCtNAt46u6gfJ5XXefC9HsGNBHtvo_Ve6oF7EXhU772aLAbXWkU1xOwQTQynixaypbRIas_kiO2MHHxXfeeL_yYZRrgtatsDBxcBRg-nUQv4TahzaGSyK42E_4YGpLa3X3Jc4t1z0SQXge7lrwlj8ysmqgO4ndlFjwPfvg0eoYqu9Qsc5Q7tazzFf9mVKMmcS1ppPutdyqNYWL62P1prw_wclP0TezW1CsypjWSVT4AuJU8YmH8nTNR1EXn8mJURLSjINv6YbZpnhBIPgUGk1JYVLcn47w
```

将 token 值输入到令牌后，单击登录即可访问 Dashboard，参考图 1-3。

图 1-3　Dashboard 页面

将 Kube-proxy 改为 ipvs 模式，因为在初始化集群的时候注释了 ipvs 配置，所以需要自行修改一下：

```
kubectl edit cm kube-proxy -n kube-system
mode: "ipvs"
```

更新 Kube-Proxy 的 Pod：

```
kubectl patch daemonset kube-proxy -p
"{\"spec\":{\"template\":{\"metadata\":{\"annotations\":{\"date\":\"`date
+'%s'`\"}}}}}" -n kube-system
```

验证 Kube-Proxy 模式

```
[root@K8S-master01 1.1.1]# curl 127.0.0.1:10249/proxyMode
ipvs
```

1.2　Kubeadm 高可用安装 K8S 集群（1.13.x 和 1.14.x）

Kubeadm 安装 Kubernetes 1.13.x 和 1.14.x 版本差异并不是很大，相对于 1.12.x 和 1.11.x 版本更加简单，只需要对其中一台 Master 初始化即可，其他 Master 节点和 Node 使用 join 即可，Master 和 Node 添加到集群中只差了一个参数，修改命令如下：

```
--experimental-control-plane
```

1.2.1　基本组件的安装

关于基本环境配置和内核升级，请参考 1.1 节。

和上节一样,需要提前安装 Kubernetes 集群的必需组件。

安装 Docker:

```
yum -y install docker-ce-17.09.1.ce-1.el7.centos
```

安装 Kubernetes 组件:

```
yum install -y kubeadm-1.13.2-0.x86_64 kubectl-1.13.2-0.x86_64 kubelet-1.13.2-0.x86_64
```

所有节点启动 Docker:

```
systemctl enable --now docker
```

配置 Kubelet:

```
[root@K8S-master01 ~]# DOCKER_CGROUPS=$(docker info | grep 'Cgroup' | cut -d' ' -f3)
[root@K8S-master01 ~]# echo $DOCKER_CGROUPS
cgroupfs
[root@K8S-master01 ~]# cat >/etc/sysconfig/kubelet<<EOF
KUBELET_EXTRA_ARGS="--cgroup-driver=$DOCKER_CGROUPS --pod-infra-container-image=registry.cn-hangzhou.aliyuncs.com/google_containers/pause-amd64:3.1"
EOF
[root@K8S-master01 ~]#
[root@K8S-master01 ~]# systemctl daemon-reload
[root@K8S-master01 ~]# systemctl enable --now kubelet
Created symlink from /etc/systemd/system/multi-user.target.wants/kubelet.service to /etc/systemd/system/kubelet.service.
```

HAProxy 和 KeepAlived 的安装请参考 1.1.4 节。

1.2.2 集群初始化

Master01 节点集群初始化和上一节演示的版本一致,但是 kubeadm-config.yaml 有所变化,去掉了内置于 Kubernetes 集群中的 Etcd 集群配置。在 1.13.x 版本中,Master02 和 Master03 无须 kubeadm-config.yaml 也可,但是为了提前下载镜像,一般也会拷贝过去。

使用 kubeadm 安装 Kubernetes 高可用集群 1.13.x 和 1.14.x 版本,kubeadm 的配置文件如下:

```
apiVersion: kubeadm.K8S.io/v1beta1
kind: ClusterConfiguration
kubernetesVersion: v1.13.2
imageRepository: registry.cn-hangzhou.aliyuncs.com/google_containers
apiServer:
  certSANs:
  - 192.168.20.10
controlPlaneEndpoint: "192.168.20.10:16443"
networking:
  # This CIDR is a Calico default. Substitute or remove for your CNI provider.
  podSubnet: "172.168.0.0/16"
---
apiVersion: kubeproxy.config.K8S.io/v1alpha1
```

```
kind: KubeProxyConfiguration
ipvs:
  minSyncPeriod: 1s
  scheduler: rr
  syncPeriod: 10s
mode: ipvs
```

和上一节不同的是直接开启了 ipvs 模式的 rr 模式，这样在初始化完成以后不用再次修改了，其中 podSubnet 为 Pod 的网段，如果安装 1.14.x，只需要将 Kubernetes 版本改成 1.14.x 即可。

Master 节点提前下载镜像：

```
kubeadm config images pull --config /root/kubeadm-config.yaml
```

Master01 节点初始化：

```
kubeadm init --config /root/kubeadm-config.yaml
```

对于 Kubernetes 1.14.x，在初始化时加入--experimental-upload-certs 参数，使集群初始化更加简单，无须再复制证书至其他节点，之后 join 时添加--certificate-key 参数即可自动加入集群。Kubernetes 1.14.x 的初始化命令如下：

```
kubeadm init --config=kubeadm-config.yaml --experimental-upload-certs
```

如果初始化失败，重置后再次初始化：

```
kubeadm reset
```

记录 token 值，在节点加入集群时使用：

```
kubeadm join 192.168.20.10:16443 --token cxwr3f.2knnb1gj83ztdg9l
--discovery-token-ca-cert-hash
sha256:41718412b5d2ccdc8b7326fd440360bf186a21dac4a0769f460ca4bdaf5d2825
```

对于 Kubernetes 1.14.x 版本，初始化完成以后生成的 Token 如下：

```
...
You can now join any number of control-plane node by running the following command
on each as a root:
    kubeadm join 192.168.0.200:6443 --token 9vr73a.a8uxyaju799qwdjv
--discovery-token-ca-cert-hash
sha256:7c2e69131a36ae2a042a339b33381c6d0d43887e2de83720eff5359e26aec866
--experimental-control-plane --certificate-key
f8902e114ef118304e561c3ecd4d0b543adc226b7a07f675f56564185ffe0c07

Please note that the certificate-key gives access to cluster sensitive data,
keep it secret!
As a safeguard, uploaded-certs will be deleted in two hours; If necessary, you
can use kubeadm init phase upload-certs to reload certs afterward.

Then you can join any number of worker nodes by running the following on each
as root:
    kubeadm join 192.168.0.200:6443 --token 9vr73a.a8uxyaju799qwdjv
--discovery-token-ca-cert-hash
sha256:7c2e69131a36ae2a042a339b33381c6d0d43887e2de83720eff5359e26aec866
```

其中，Master 节点使用--experimental-control-plane 和--certificate-key 参数即可完成初始化，并

以 Master 的角色加入集群：

```
    kubeadm join 192.168.0.200:6443 --token 9vr73a.a8uxyaju799qwdjv
--discovery-token-ca-cert-hash
sha256:7c2e69131a36ae2a042a339b33381c6d0d43887e2de83720eff5359e26aec866
--experimental-control-plane --certificate-key
f8902e114ef118304e561c3ecd4d0b543adc226b7a07f675f56564185ffe0c07
```

所有 Master 节点配置环境变量：

```
cat <<EOF >> /root/.bashrc
export KUBECONFIG=/etc/kubernetes/admin.conf
EOF
source /root/.bashrc
```

查看节点状态：

```
[root@K8S-master01 ~]# kubectl get nodes
NAME            STATUS     ROLES    AGE     VERSION
K8S-master01    NotReady   master   2m11s   v1.13.2
```

查看 Pod 状态：

```
[root@K8S-master01 ~]# kubectl get pods -n kube-system -o wide
  NAME                                        READY    STATUS              RESTARTS     AGE
IP              NODE              NOMINATED NODE    READINESS GATES
  coredns-89cc84847-2h7r6                     0/1      ContainerCreating   0
3m12s  <none>          K8S-master01      <none>            <none>
  coredns-89cc84847-fhwbr                     0/1      ContainerCreating   0
3m12s  <none>          K8S-master01      <none>            <none>
  etcd-K8S-master01                           1/1      Running             0            2m31s
192.168.20.20   K8S-master01      <none>            <none>
  kube-apiserver-K8S-master01                 1/1      Running             0
2m36s  192.168.20.20   K8S-master01      <none>            <none>
  kube-controller-manager-K8S-master01        1/1      Running             0
2m39s  192.168.20.20   K8S-master01      <none>            <none>
  kube-proxy-kb95s                            1/1      Running             0            3m12s
192.168.20.20   K8S-master01      <none>            <none>
  kube-scheduler-K8S-master01                 1/1      Running             0
2m46s  192.168.20.20   K8S-master01      <none>            <none>
```

1.2.3　Calico 组件的安装

可安装截止本书截稿时的最新版 3.6.1，也可以参考 1.1.5 节，POD_CIDR 为上述配置的 podSubnet：

```
POD_CIDR="<your-pod-cidr>" \
sed -i -e "s?192.168.0.0/16?$POD_CIDR?g" calico/v3.6.1/calico.yaml
kubectl apply -f calico/v3.6.1/calico.yaml
```

1.2.4　高可用 Master

Kubernetes 1.13.x 版需要复制证书至其他 Master 节点，1.14.x 版则无须再复制证书至其他

Master 节点：

```
USER=root
CONTROL_PLANE_IPS="K8S-master02 K8S-master03"
for host in $CONTROL_PLANE_IPS; do
    ssh "${USER}"@$host "mkdir -p /etc/kubernetes/pki/etcd"
    scp /etc/kubernetes/pki/ca.crt "${USER}"@$host:/etc/kubernetes/pki/ca.crt
    scp /etc/kubernetes/pki/ca.key "${USER}"@$host:/etc/kubernetes/pki/ca.key
    scp /etc/kubernetes/pki/sa.key "${USER}"@$host:/etc/kubernetes/pki/sa.key
    scp /etc/kubernetes/pki/sa.pub "${USER}"@$host:/etc/kubernetes/pki/sa.pub
    scp /etc/kubernetes/pki/front-proxy-ca.crt "${USER}"@$host:/etc/kubernetes/pki/front-proxy-ca.crt
    scp /etc/kubernetes/pki/front-proxy-ca.key "${USER}"@$host:/etc/kubernetes/pki/front-proxy-ca.key
    scp /etc/kubernetes/pki/etcd/ca.crt "${USER}"@$host:/etc/kubernetes/pki/etcd/ca.crt
    scp /etc/kubernetes/pki/etcd/ca.key "${USER}"@$host:/etc/kubernetes/pki/etcd/ca.key
    scp /etc/kubernetes/admin.conf "${USER}"@$host:/etc/kubernetes/admin.conf
done
```

Master02 提前下载镜像：

```
kubeadm config images pull --config /root/kubeadm-config.yaml
```

Master02 加入集群，与 Node 节点相差的参数就是 --experimental-control-plane：

```
kubeadm join 192.168.20.10:16443 --token cxwr3f.2knnb1gj83ztdg9l --discovery-token-ca-cert-hash sha256:41718412b5d2ccdc8b7326fd440360bf186a21dac4a0769f460ca4bdaf5d2825 --experimental-control-plane
```

对于 Kubernetes 1.14.x，使用如下命令加入集群，多了一个 --certificate-key 参数：

```
kubeadm join 192.168.0.200:6443 --token 9vr73a.a8uxyaju799qwdjv --discovery-token-ca-cert-hash sha256:7c2e69131a36ae2a042a339b33381c6d0d43887e2de83720eff5359e26aec866 --experimental-control-plane --certificate-key f8902e114ef118304e561c3ecd4d0b543adc226b7a07f675f56564185ffe0c07
```

反馈如下：

```
......

This node has joined the cluster and a new control plane instance was created:

* Certificate signing request was sent to apiserver and approval was received.
* The Kubelet was informed of the new secure connection details.
* Master label and taint were applied to the new node.
* The Kubernetes control plane instances scaled up.
* A new etcd member was added to the local/stacked etcd cluster.
```

```
To start administering your cluster from this node, you need to run the following
as a regular user:

    mkdir -p $HOME/.kube
    sudo cp -i /etc/kubernetes/admin.conf $HOME/.kube/config
    sudo chown $(id -u):$(id -g) $HOME/.kube/config

Run 'kubectl get nodes' to see this node join the cluster.
```

Master01 查看状态：

```
[root@K8S-master01 K8S-ha-install]# kubectl get no
NAME            STATUS   ROLES    AGE    VERSION
K8S-master01    Ready    master   15m    v1.13.2
K8S-master02    Ready    master   9m55s  v1.13.2
```

其他 Master 节点操作相同，查看 Master 最终的状态：

```
[root@K8S-master01 ~]# ipvsadm -ln
IP Virtual Server version 1.2.1 (size=4096)
Prot LocalAddress:Port Scheduler Flags
  -> RemoteAddress:Port           Forward Weight ActiveConn InActConn
TCP  10.96.0.1:443 rr
  -> 192.168.20.20:6443           Masq    1      4          0
  -> 192.168.20.21:6443           Masq    1      0          0
  -> 192.168.20.22:6443           Masq    1      0          0
TCP  10.96.0.10:53 rr
  -> 172.168.0.10:53              Masq    1      0          0
  -> 172.168.0.11:53              Masq    1      0          0
TCP  10.102.221.48:5473 rr
UDP  10.96.0.10:53 rr
  -> 172.168.0.10:53              Masq    1      0          0
  -> 172.168.0.11:53              Masq    1      0          0
[root@K8S-master01 ~]# kubectl get po -n kube-system
NAME                                        READY   STATUS    RESTARTS   AGE
calico-node-49dwr                           2/2     Running   0          26m
calico-node-kz2d4                           2/2     Running   0          22m
calico-node-zwnmq                           2/2     Running   0          4m6s
coredns-89cc84847-dgxlw                     1/1     Running   0          27m
coredns-89cc84847-n77x6                     1/1     Running   0          27m
etcd-K8S-master01                           1/1     Running   0          27m
etcd-K8S-master02                           1/1     Running   0          22m
etcd-K8S-master03                           1/1     Running   0          4m5s
kube-apiserver-K8S-master01                 1/1     Running   0          27m
kube-apiserver-K8S-master02                 1/1     Running   0          22m
kube-apiserver-K8S-master03                 1/1     Running   3          4m6s
kube-controller-manager-K8S-master01        1/1     Running   1          27m
kube-controller-manager-K8S-master02        1/1     Running   0          22m
kube-controller-manager-K8S-master03        1/1     Running   0          4m6s
kube-proxy-f9qc5                            1/1     Running   0          27m
kube-proxy-k55bg                            1/1     Running   0          22m
kube-proxy-kbg9c                            1/1     Running   0          4m6s
kube-scheduler-K8S-master01                 1/1     Running   1          27m
kube-scheduler-K8S-master02                 1/1     Running   0          22m
kube-scheduler-K8S-master03                 1/1     Running   0          4m6s
[root@K8S-master01 ~]# kubectl get no
```

```
NAME           STATUS   ROLES    AGE     VERSION
K8S-master01   Ready    master   28m     v1.13.2
K8S-master02   Ready    master   22m     v1.13.2
K8S-master03   Ready    master   4m16s   v1.13.2
```

查看 CSR：

```
[root@K8S-master01 ~]# kubectl get csr
NAME                                                   AGE    REQUESTOR
CONDITION
  csr-6mqbv                                            28m
system:node:K8S-master01    Approved,Issued
  node-csr-GPLcR1G4Nchf-zuB5DaTWncoluMuENUfKvWKs0j2GdQ  23m
system:bootstrap:9zp70m     Approved,Issued
  node-csr-cxAxrkllyidkBuZ8fck6fwq-ht1_u6s0snbDErM8bIs  4m51s
system:bootstrap:9zp70m     Approved,Issued
```

在所有 Master 节点上允许 HPA 采集数据，修改后自动重启：

```
vi /etc/kubernetes/manifests/kube-controller-manager.yaml
  - --horizontal-pod-autoscaler-use-rest-clients=false
```

1.2.5　Node 节点的配置

在 1.13.x 和 1.14.x 版本中，Node 节点和 Master 节点加入集群的方式与 1.11.x 和 1.12.x 版本相比只是少了--experimental-control-plane 参数。

将 Node 节点加入集群，所有 Node 节点配置相同：

```
kubeadm join 192.168.20.10:16443 --token cxwr3f.2knnb1gj83ztdg9l
--discovery-token-ca-cert-hash
sha256:41718412b5d2ccdc8b7326fd440360bf186a21dac4a0769f460ca4bdaf5d2825
```

反馈如下：

```
......
[kubelet-start] Activating the kubelet service
[tlsbootstrap] Waiting for the kubelet to perform the TLS Bootstrap...
[patchnode] Uploading the CRI Socket information "/var/run/dockershim.sock"
to the Node API object "K8S-node02" as an annotation

This node has joined the cluster:
* Certificate signing request was sent to apiserver and a response was received.
* The Kubelet was informed of the new secure connection details.

Run 'kubectl get nodes' on the master to see this node join the cluster.
```

查看 Master 节点的状态：

```
[root@K8S-master01 K8S-ha-install]# kubectl get po -n kube-system
NAME                READY   STATUS    RESTARTS   AGE
calico-node-49dwr   2/2     Running   0          13h
calico-node-9nmhb   2/2     Running   0          11m
calico-node-k5nmt   2/2     Running   0          11m
calico-node-kz2d4   2/2     Running   0          13h
calico-node-zwnmq   2/2     Running   0          13h
```

```
coredns-89cc84847-dgxlw                         1/1     Running   0   13h
coredns-89cc84847-n77x6                         1/1     Running   0   13h
etcd-K8S-master01                               1/1     Running   0   13h
etcd-K8S-master02                               1/1     Running   0   13h
etcd-K8S-master03                               1/1     Running   0   13h
kube-apiserver-K8S-master01                     1/1     Running   0   18m
kube-apiserver-K8S-master02                     1/1     Running   0   17m
kube-apiserver-K8S-master03                     1/1     Running   0   16m
kube-controller-manager-K8S-master01   1/1      Running   0   19m
kube-controller-manager-K8S-master02   1/1      Running   1   19m
kube-controller-manager-K8S-master03   1/1      Running   0   19m
kube-proxy-cl2zv                                1/1     Running   0   11m
kube-proxy-f9qc5                                1/1     Running   0   13h
kube-proxy-hkcq5                                1/1     Running   0   11m
kube-proxy-k55bg                                1/1     Running   0   13h
kube-proxy-kbg9c                                1/1     Running   0   13h
kube-scheduler-K8S-master01                     1/1     Running   1   13h
kube-scheduler-K8S-master02                     1/1     Running   0   13h
kube-scheduler-K8S-master03                     1/1     Running   0   13h
You have new mail in /var/spool/mail/root
[root@K8S-master01 K8S-ha-install]# kubectl get no
NAME            STATUS   ROLES    AGE   VERSION
K8S-master01    Ready    master   13h   v1.13.2
K8S-master02    Ready    master   13h   v1.13.2
K8S-master03    Ready    master   13h   v1.13.2
K8S-node01      Ready    <none>   11m   v1.13.2
K8S-node02      Ready    <none>   11m   v1.13.2
```

关于 Metrics 和 Dashboard 的部署请参考 1.1.8 节和 1.1.9 节。

1.3 二进制高可用安装 K8S 集群（1.13.x 和 1.14.x）

上一节讲解了使用 Kubeadm 安装高可用 Kubernetes 集群，虽然现在 kubeadm 是官方默认的安装方式，但是在生产环境中仍然不建议使用 Kubeadm 安装方式。在实际测试中，二进制安装方式比 Kubeadm 安装方式更加稳定可靠，并且集群的恢复能力比 Kubeadm 要高。不过在线下的测试环境，为了能够快速实现测试及部署可以使用 Kubeadm 安装方式，等到在生产环境中时仍然建议采用二进制安装方式。

本节介绍 Kubernetes 1.13.x 和 1.14.x 版本的高可用集群的安装，在二进制安装方式下，很多步骤需要自己手动完成，比如证书和配置文件的生成等，在二进制安装过程中，其他版本的安装过程基本一致，替换二进制文件的版本即可。

关于基本环境的配置请参考 1.1.1 节（yum 仓库配置可省略），同样采用 5 台主机，3 台 Master 和 2 台 Node。

> **注　意**
>
> 与之前不同的是，本例的 VIP 为 192.168.20.110

1.3.1　基本组件安装

关于内核升级，请参考 1.1.2 节。

和 Kubeadm 安装方式一致，同样需要提前安装集群中必需的组件。

所有节点安装 Docker：

```
# 基本工具安装
yum install -y yum-utils device-mapper-persistent-data lvm2
# Docker 源配置
yum-config-manager --add-repo http://mirrors.aliyun.com/docker-ce/linux/centos/docker-ce.repo
  yum makecache fast
# 查看可用 Docker 版本
yum list docker-ce.x86_64 --showduplicates | sort -r
# 安装指定版本 Docker，按需修改
yum -y install docker-ce-17.09.1.ce-1.el7.centos
```

所有节点开启 Docker 并设置为开机自启动：

```
systemctl enable --now docker
```

下载 Kubernetes，本例安装的是 13.1，其他版本请自行修改：

```
    wget https://storage.googleapis.com/kubernetes-release/release/13.1/kubernetes-server-linux-amd64.tar.gz
    tar -xf kubernetes-server-linux-amd64.tar.gz --strip-components=3 -C /usr/local/bin kubernetes/server/bin/kube{let,ctl,-apiserver,-controller-manager,-scheduler,-proxy}
```

也可在 GitHub 上下载：

```
    https://github.com/kubernetes/kubernetes/blob/master/CHANGELOG-1.13.md#downloads-for-v1131
```

下载 Etcd 文件，如果安装的是 1.14.x 版，则可以选择安装 Etcd 的 3.3.10 版：

```
[root@K8S-master01 packages]# wget https://github.com/etcd-io/etcd/releases/download/v3.3.9/etcd-v3.3.9-linux-amd64.tar.gz

[root@K8S-master01 packages]# tar -zxvf etcd-v3.3.9-linux-amd64.tar.gz --strip-components=1 -C /usr/local/bin etcd-v3.3.9-linux-amd64/etcd{,ctl}
    etcd-v3.3.9-linux-amd64/etcdctl
    etcd-v3.3.9-linux-amd64/etcd

[root@K8S-master01 packages]# etcd --version
    etcd Version: 3.3.9
```

```
Git SHA: fca8add78
Go Version: go1.10.3
Go OS/Arch: linux/amd64
```

将各组件分发至其他节点：

```
MasterNodes='K8S-master02 K8S-master03'
WorkNodes='K8S-node01 K8S-node02'

# Master
for NODE in $MasterNodes; do
    echo $NODE
    scp /usr/local/bin/kube{let,ctl,-apiserver,-controller-manager,-scheduler,-proxy} $NODE:/usr/local/bin/;
    scp /usr/local/bin/etcd* $NODE:/usr/local/bin/
done

#Work
for NODE in $WorkNodes; do
    scp /usr/local/bin/kube{let,-proxy} $NODE:/usr/local/bin/
done
```

所有 Master 节点安装 HAProxy 和 KeepAlived，可参考 1.1.4 节

1.3.2　CNI 安装

CNI（Container Network Interface，容器网络接口）是 CNCF 旗下的一个项目，由一组用于配置容器的网络接口的规范和库组成。CNI 主要用于解决容器网络互联的配置并支持多种网络模式。CNI 的安装步骤如下。

所有节点创建 CNI 目录：

```
mkdir -p /opt/cni/bin
```

Master01 下载 CNI，如果安装的是 1.14.x 版本，可以安装 CNI 的 0.7.5 版本：

```
wget https://github.com/containernetworking/plugins/releases/download/v0.7.1/cni-plugins-amd64-v0.7.1.tgz

tar -zxf cni-plugins-amd64-v0.7.1.tgz -C /opt/cni/bin
```

将 CNI 分发至其他节点：

```
# 分发 CNI 文件到 Master
for NODE in $MasterNodes; do
    ssh $NODE 'mkdir -p /opt/cni/bin'
    scp /opt/cni/bin/* $NODE:/opt/cni/bin/
done
# 分发到 Node
for NODE in $WorkNodes; do
    ssh $NODE 'mkdir -p /opt/cni/bin'
    scp /opt/cni/bin/* $NODE:/opt/cni/bin/
done
```

1.3.3 生成证书

在 Kubeadm 安装方式下，初始化时会自动生成证书，但在二进制安装方式下，需要手动生成证书，可以使用 OpenSSL 或者 cfssl。具体操作步骤如下：

Master01 安装 cfssl：

```
wget "https://pkg.cfssl.org/R1.2/cfssl_linux-amd64" -O /usr/local/bin/cfssl
wget "https://pkg.cfssl.org/R1.2/cfssljson_linux-amd64" -O /usr/local/bin/cfssljson
chmod +x /usr/local/bin/cfssl /usr/local/bin/cfssljson
```

所有 Master 节点创建 Etcd 证书目录：

```
mkdir /etc/etcd/ssl -p
```

Master01 生成 Etcd 证书：

```
cd chap01/1.3/pki/
# 使用从 CSR 文件生成 etcd 的 CA key 和 Certificate
[root@K8S-master01 pki]# cfssl gencert -initca etcd-ca-csr.json | cfssljson -bare /etc/etcd/ssl/etcd-ca
2019/01/02 18:17:53 [INFO] generating a new CA key and certificate from CSR
2019/01/02 18:17:53 [INFO] generate received request
2019/01/02 18:17:53 [INFO] received CSR
2019/01/02 18:17:53 [INFO] generating key: rsa-2048
2019/01/02 18:17:54 [INFO] encoded CSR
2019/01/02 18:17:54 [INFO] signed certificate with serial number 176500636363716542593556506106786978367462282402

# 生成证书
cfssl gencert \
  -ca=/etc/etcd/ssl/etcd-ca.pem \
  -ca-key=/etc/etcd/ssl/etcd-ca-key.pem \
  -config=ca-config.json \
  -hostname=127.0.0.1,K8S-master01,K8S-master02,K8S-master03,192.168.20.20,192.168.20.21,192.168.20.22 \
  -profile=kubernetes \
  etcd-csr.json | cfssljson -bare /etc/etcd/ssl/etcd
```

将证书复制到其他节点，当前 Etcd 集群部署在 Master 节点上，在大规模集群环境中建议部署在集群之外，并且使用 SSD 硬盘作为 Etcd 的存储：

```
for NODE in $MasterNodes; do
    ssh $NODE "mkdir -p /etc/etcd/ssl"
    for FILE in etcd-ca-key.pem etcd-ca.pem etcd-key.pem etcd.pem; do
      scp /etc/etcd/ssl/${FILE} $NODE:/etc/etcd/ssl/${FILE}
    done
done
```

生成 Kubernetes CA：

```
[root@K8S-master01 pki] mkdir -p /etc/kubernetes/pki
[root@K8S-master01 pki]# cfssl gencert -initca ca-csr.json | cfssljson -bare /etc/kubernetes/pki/ca
2019/01/02 11:23:49 [INFO] generating a new CA key and certificate from CSR
```

```
2019/01/02 11:23:49 [INFO] generate received request
2019/01/02 11:23:49 [INFO] received CSR
2019/01/02 11:23:49 [INFO] generating key: rsa-2048
2019/01/02 11:23:49 [INFO] encoded CSR
2019/01/02 11:23:49 [INFO] signed certificate with serial number
241147988728833611135804600660533145120543777046
```

生成 API Server 证书，10.96.0.1 是 Cluster IP 的 Kubernetes 端点，用于集群里的 Pod 调用 K8S 的 API Server，使用时注意不要和公司网络在同一个网段：

```
cfssl gencert   -ca=/etc/kubernetes/pki/ca.pem
-ca-key=/etc/kubernetes/pki/ca-key.pem   -config=ca-config.json
-hostname=10.96.0.1,192.168.20.110,127.0.0.1,kubernetes,kubernetes.default,kubernetes.default.svc,kubernetes.default.svc.cluster,kubernetes.default.svc.cluster.local,192.168.20.20,192.168.20.21,192.168.20.22   -profile=kubernetes
apiserver-csr.json | cfssljson -bare /etc/kubernetes/pki/apiserver
```

创建 Front Proxy 证书：

```
cfssl gencert   -initca front-proxy-ca-csr.json | cfssljson -bare
/etc/kubernetes/pki/front-proxy-ca

cfssl gencert   -ca=/etc/kubernetes/pki/front-proxy-ca.pem
-ca-key=/etc/kubernetes/pki/front-proxy-ca-key.pem   -config=ca-config.json
-profile=kubernetes   front-proxy-client-csr.json | cfssljson -bare
/etc/kubernetes/pki/front-proxy-client
```

生成 ControllerManager 证书：

```
cfssl gencert \
  -ca=/etc/kubernetes/pki/ca.pem \
  -ca-key=/etc/kubernetes/pki/ca-key.pem \
  -config=ca-config.json \
  -profile=kubernetes \
  manager-csr.json | cfssljson -bare /etc/kubernetes/pki/controller-manager
```

创建 ControllerManager 的 kubeconfig 文件，注意修改--server 的地址：

```
kubectl config set-cluster kubernetes \
    --certificate-authority=/etc/kubernetes/pki/ca.pem \
    --embed-certs=true \
    --server=https://192.168.20.110:8443 \
    --kubeconfig=/etc/kubernetes/controller-manager.kubeconfig

kubectl config set-credentials system:kube-controller-manager \
    --client-certificate=/etc/kubernetes/pki/controller-manager.pem \
    --client-key=/etc/kubernetes/pki/controller-manager-key.pem \
    --embed-certs=true \
    --kubeconfig=/etc/kubernetes/controller-manager.kubeconfig

kubectl config set-context system:kube-controller-manager@kubernetes \
    --cluster=kubernetes \
    --user=system:kube-controller-manager \
    --kubeconfig=/etc/kubernetes/controller-manager.kubeconfig

kubectl config use-context system:kube-controller-manager@kubernetes \
```

```
  --kubeconfig=/etc/kubernetes/controller-manager.kubeconfig
```

生成 Scheduler 证书：

```
cfssl gencert \
  -ca=/etc/kubernetes/pki/ca.pem \
  -ca-key=/etc/kubernetes/pki/ca-key.pem \
  -config=ca-config.json \
  -profile=kubernetes \
  scheduler-csr.json | cfssljson -bare /etc/kubernetes/pki/scheduler
```

生成 Scheduler 的 kubeconfig 文件，注意修改--server 的地址：

```
kubectl config set-cluster kubernetes \
    --certificate-authority=/etc/kubernetes/pki/ca.pem \
    --embed-certs=true \
    --server=https://192.168.20.110:8443 \
    --kubeconfig=/etc/kubernetes/scheduler.kubeconfig

kubectl config set-credentials system:kube-scheduler \
    --client-certificate=/etc/kubernetes/pki/scheduler.pem \
    --client-key=/etc/kubernetes/pki/scheduler-key.pem \
    --embed-certs=true \
    --kubeconfig=/etc/kubernetes/scheduler.kubeconfig

kubectl config set-context system:kube-scheduler@kubernetes \
    --cluster=kubernetes \
    --user=system:kube-scheduler \
    --kubeconfig=/etc/kubernetes/scheduler.kubeconfig

kubectl config use-context system:kube-scheduler@kubernetes \
    --kubeconfig=/etc/kubernetes/scheduler.kubeconfig
```

生成 Admin Certificate：

```
cfssl gencert \
  -ca=/etc/kubernetes/pki/ca.pem \
  -ca-key=/etc/kubernetes/pki/ca-key.pem \
  -config=ca-config.json \
  -profile=kubernetes \
  admin-csr.json | cfssljson -bare /etc/kubernetes/pki/admin
```

生成 Admin 的 kubeconfig 文件，注意修改--server 的地址：

```
   kubectl config set-cluster kubernetes
--certificate-authority=/etc/kubernetes/pki/ca.pem     --embed-certs=true
--server=https://192.168.20.110:8443
--kubeconfig=/etc/kubernetes/admin.kubeconfig

   kubectl config set-credentials kubernetes-admin
--client-certificate=/etc/kubernetes/pki/admin.pem
--client-key=/etc/kubernetes/pki/admin-key.pem     --embed-certs=true
--kubeconfig=/etc/kubernetes/admin.kubeconfig

   kubectl config set-context kubernetes-admin@kubernetes
--cluster=kubernetes     --user=kubernetes-admin
--kubeconfig=/etc/kubernetes/admin.kubeconfig
```

```
kubectl config use-context kubernetes-admin@kubernetes
--kubeconfig=/etc/kubernetes/admin.kubeconfig
```

生成所有 Master 节点的 Kubelet 凭证：

```
for NODE in K8S-master01 K8S-master02 K8S-master03; do
    \cp kubelet-csr.json kubelet-$NODE-csr.json;
    sed -i "s/\$NODE/$NODE/g" kubelet-$NODE-csr.json;
    cfssl gencert \
      -ca=/etc/kubernetes/pki/ca.pem \
      -ca-key=/etc/kubernetes/pki/ca-key.pem \
      -config=ca-config.json \
      -hostname=$NODE \
      -profile=kubernetes \
      kubelet-$NODE-csr.json | cfssljson -bare /etc/kubernetes/pki/kubelet-$NODE;
    rm -f kubelet-$NODE-csr.json
  done
```

复制证书到其他节点：

```
for NODE in K8S-master01 K8S-master02 K8S-master03; do
    ssh $NODE "mkdir -p /etc/kubernetes/pki"
    scp /etc/kubernetes/pki/ca.pem $NODE:/etc/kubernetes/pki/ca.pem
    scp /etc/kubernetes/pki/kubelet-$NODE-key.pem $NODE:/etc/kubernetes/pki/kubelet-key.pem
    scp /etc/kubernetes/pki/kubelet-$NODE.pem $NODE:/etc/kubernetes/pki/kubelet.pem
    rm -f /etc/kubernetes/pki/kubelet-$NODE-key.pem /etc/kubernetes/pki/kubelet-$NODE.pem
  done
```

生成所有 Master 节点的 kubeconfig 文件，注意修改--server 的地址：

```
for NODE in K8S-master01 K8S-master02 K8S-master03; do
    ssh $NODE "cd /etc/kubernetes/pki && \
      kubectl config set-cluster kubernetes \
        --certificate-authority=/etc/kubernetes/pki/ca.pem \
        --embed-certs=true \
        --server=https://192.168.20.110:8443 \
        --kubeconfig=/etc/kubernetes/kubelet.kubeconfig && \
      kubectl config set-credentials system:node:${NODE} \
        --client-certificate=/etc/kubernetes/pki/kubelet.pem \
        --client-key=/etc/kubernetes/pki/kubelet-key.pem \
        --embed-certs=true \
        --kubeconfig=/etc/kubernetes/kubelet.kubeconfig && \
      kubectl config set-context system:node:${NODE}@kubernetes \
        --cluster=kubernetes \
        --user=system:node:${NODE} \
        --kubeconfig=/etc/kubernetes/kubelet.kubeconfig && \
      kubectl config use-context system:node:${NODE}@kubernetes \
        --kubeconfig=/etc/kubernetes/kubelet.kubeconfig"
  done
```

创建 ServiceAccount Key：

```
[root@K8S-master01 pki]# openssl genrsa -out /etc/kubernetes/pki/sa.key 2048
Generating RSA private key, 2048 bit long modulus
.......................+++
............+++
e is 65537 (0x10001)
[root@K8S-master01 pki]# openssl rsa -in /etc/kubernetes/pki/sa.key -pubout
-out /etc/kubernetes/pki/sa.pub
writing RSA key
[root@K8S-master01 pki]# ls /etc/kubernetes/pki/sa.*
/etc/kubernetes/pki/sa.key   /etc/kubernetes/pki/sa.pub
```

复制到其他节点:

```
for NODE in K8S-master02 K8S-master03; do
    for FILE in $(ls /etc/kubernetes/pki | grep -v etcd); do
      scp /etc/kubernetes/pki/${FILE} $NODE:/etc/kubernetes/pki/${FILE}
    done
    for FILE in admin.kubeconfig controller-manager.kubeconfig
scheduler.kubeconfig; do
      scp /etc/kubernetes/${FILE} $NODE:/etc/kubernetes/${FILE}
    done
done
```

1.3.4 系统组件配置

在二进制安装方式下，Kubernetes 的组件都是以守护进程的方式运行在宿主机上，相比于 Kubeadm 安装方式，虽然配置过程较复杂，但是程序运行较稳定，并且恢复能力较强。

首先配置高可用 Etcd 集群。在 Master01 节点上创建 etcd-master01 的配置文件，注意修改对应的 IP 地址和 name（名字）：

```
[root@K8S-master02 bin]# cat /etc/etcd/etcd.config.yml

name: 'K8S-master01'
data-dir: /var/lib/etcd
wal-dir: /var/lib/etcd/wal
snapshot-count: 5000
heartbeat-interval: 100
election-timeout: 1000
quota-backend-bytes: 0
listen-peer-urls: 'https://192.168.20.20:2380'
listen-client-urls: 'https://192.168.20.20:2379,http://127.0.0.1:2379'
max-snapshots: 3
max-wals: 5
cors:
initial-advertise-peer-urls: 'https://192.168.20.20:2380'
advertise-client-urls: 'https://192.168.20.20:2379'
discovery:
discovery-fallback: 'proxy'
discovery-proxy:
discovery-srv:
initial-cluster: 'K8S-master01=https://192.168.20.20:2380,K8S-master02=
https://192.168.20.21:2380,K8S-master03=https://192.168.20.22:2380'
    initial-cluster-token: 'etcd-K8S-cluster'
```

```yaml
    initial-cluster-state: 'new'
    strict-reconfig-check: false
    enable-v2: true
    enable-pprof: true
    proxy: 'off'
    proxy-failure-wait: 5000
    proxy-refresh-interval: 30000
    proxy-dial-timeout: 1000
    proxy-write-timeout: 5000
    proxy-read-timeout: 0
    client-transport-security:
      ca-file: '/etc/kubernetes/pki/etcd/etcd-ca.pem'
      cert-file: '/etc/kubernetes/pki/etcd/etcd.pem'
      key-file: '/etc/kubernetes/pki/etcd/etcd-key.pem'
      client-cert-auth: true
      trusted-ca-file: '/etc/kubernetes/pki/etcd/etcd-ca.pem'
      auto-tls: true
    peer-transport-security:
      ca-file: '/etc/kubernetes/pki/etcd/etcd-ca.pem'
      cert-file: '/etc/kubernetes/pki/etcd/etcd.pem'
      key-file: '/etc/kubernetes/pki/etcd/etcd-key.pem'
      peer-client-cert-auth: true
      trusted-ca-file: '/etc/kubernetes/pki/etcd/etcd-ca.pem'
      auto-tls: true
    debug: false
    log-package-levels:
    log-output: default
    force-new-cluster: false
```

etcd-master02 配置文件：

```yaml
    [root@K8S-master02 bin]# cat /etc/etcd/etcd.config.yml
    name: 'K8S-master02'
    data-dir: /var/lib/etcd
    wal-dir: /var/lib/etcd/wal
    snapshot-count: 5000
    heartbeat-interval: 100
    election-timeout: 1000
    quota-backend-bytes: 0
    listen-peer-urls: 'https://192.168.20.21:2380'
    listen-client-urls: 'https://192.168.20.21:2379,http://127.0.0.1:2379'
    max-snapshots: 3
    max-wals: 5
    cors:
    initial-advertise-peer-urls: 'https://192.168.20.21:2380'
    advertise-client-urls: 'https://192.168.20.21:2379'
    discovery:
    discovery-fallback: 'proxy'
    discovery-proxy:
    discovery-srv:
    initial-cluster: 'K8S-master01=https://192.168.20.20:2380,K8S-master02=https://192.168.20.21:2380,K8S-master03=https://192.168.20.22:2380'
    initial-cluster-token: 'etcd-K8S-cluster'
    initial-cluster-state: 'new'
    strict-reconfig-check: false
```

```
    enable-v2: true
    enable-pprof: true
    proxy: 'off'
    proxy-failure-wait: 5000
    proxy-refresh-interval: 30000
    proxy-dial-timeout: 1000
    proxy-write-timeout: 5000
    proxy-read-timeout: 0
    client-transport-security:
      ca-file: '/etc/kubernetes/pki/etcd/etcd-ca.pem'
      cert-file: '/etc/kubernetes/pki/etcd/etcd.pem'
      key-file: '/etc/kubernetes/pki/etcd/etcd-key.pem'
      client-cert-auth: true
      trusted-ca-file: '/etc/kubernetes/pki/etcd/etcd-ca.pem'
      auto-tls: true
    peer-transport-security:
      ca-file: '/etc/kubernetes/pki/etcd/etcd-ca.pem'
      cert-file: '/etc/kubernetes/pki/etcd/etcd.pem'
      key-file: '/etc/kubernetes/pki/etcd/etcd-key.pem'
      peer-client-cert-auth: true
      trusted-ca-file: '/etc/kubernetes/pki/etcd/etcd-ca.pem'
      auto-tls: true
    debug: false
    log-package-levels:
    log-output: default
    force-new-cluster: false
```

etcd-master03 配置文件：

```
[root@K8S-master03 bin]# cat /etc/etcd/etcd.config.yml
name: 'K8S-master03'
data-dir: /var/lib/etcd
wal-dir: /var/lib/etcd/wal
snapshot-count: 5000
heartbeat-interval: 100
election-timeout: 1000
quota-backend-bytes: 0
listen-peer-urls: 'https://192.168.20.22:2380'
listen-client-urls: 'https://192.168.20.22:2379,http://127.0.0.1:2379'
max-snapshots: 3
max-wals: 5
cors:
initial-advertise-peer-urls: 'https://192.168.20.22:2380'
advertise-client-urls: 'https://192.168.20.22:2379'
discovery:
discovery-fallback: 'proxy'
discovery-proxy:
discovery-srv:
initial-cluster: 'K8S-master01=https://192.168.20.20:2380,K8S-master02=https://192.168.20.21:2380,K8S-master03=https://192.168.20.22:2380'
    initial-cluster-token: 'etcd-K8S-cluster'
    initial-cluster-state: 'new'
    strict-reconfig-check: false
    enable-v2: true
    enable-pprof: true
```

```
  proxy: 'off'
  proxy-failure-wait: 5000
  proxy-refresh-interval: 30000
  proxy-dial-timeout: 1000
  proxy-write-timeout: 5000
  proxy-read-timeout: 0
  client-transport-security:
    ca-file: '/etc/kubernetes/pki/etcd/etcd-ca.pem'
    cert-file: '/etc/kubernetes/pki/etcd/etcd.pem'
    key-file: '/etc/kubernetes/pki/etcd/etcd-key.pem'
    client-cert-auth: true
    trusted-ca-file: '/etc/kubernetes/pki/etcd/etcd-ca.pem'
    auto-tls: true
  peer-transport-security:
    ca-file: '/etc/kubernetes/pki/etcd/etcd-ca.pem'
    cert-file: '/etc/kubernetes/pki/etcd/etcd.pem'
    key-file: '/etc/kubernetes/pki/etcd/etcd-key.pem'
    peer-client-cert-auth: true
    trusted-ca-file: '/etc/kubernetes/pki/etcd/etcd-ca.pem'
    auto-tls: true
  debug: false
  log-package-levels:
  log-output: default
  force-new-cluster: false
```

所有 Master 节点配置 etcd.service：

```
[root@K8S-master03 bin]# cat /usr/lib/systemd/system/etcd.service
[Unit]
Description=Etcd Service
Documentation=https://coreos.com/etcd/docs/latest/
After=network.target

[Service]
Type=notify
ExecStart=/usr/local/bin/etcd --config-file=/etc/etcd/etcd.config.yml
Restart=on-failure
RestartSec=10
LimitNOFILE=65536

[Install]
WantedBy=multi-user.target
Alias=etcd3.service
```

所有 Master 节点启动 Etcd：

```
mkdir /etc/kubernetes/pki/etcd
ln -s /etc/etcd/ssl/* /etc/kubernetes/pki/etcd/
systemctl daemon-reload
systemctl enable --now etcd
```

查看状态：

```
[root@K8S-master03 ~]# systemctl status etcd
  etcd.service - Etcd Service
  Loaded: loaded (/usr/lib/systemd/system/etcd.service; enabled; vendor
```

```
preset: disabled)
   Active: active (running) since Mon 2019-02-18 19:41:08 CST; 46s ago
     Docs: https://coreos.com/etcd/docs/latest/
 Main PID: 17190 (etcd)
   CGroup: /system.slice/etcd.service
           └─17190 /usr/local/bin/etcd --config-file=/etc/etcd/etcd.config.yml


......
   [root@K8S-master01 master]# etcdctl     --cert-file /etc/etcd/ssl/etcd.pem
--key-file /etc/etcd/ssl/etcd-key.pem      --ca-file /etc/etcd/ssl/etcd-ca.pem
--endpoints
https://192.168.20.20:2379,https://192.168.20.21:2379,https://192.168.20.22:23
79 cluster-health

   member 11dee1a95949e017 is healthy: got healthy result from
https://192.168.20.20:2379
   member 62f054c5331e9fc6 is healthy: got healthy result from
https://192.168.20.22:2379
   member 8a71ace3147f5322 is healthy: got healthy result from
https://192.168.20.21:2379
   cluster is healthy
```

本例高可用配置同样使用的是 HAProxy 和 KeepAlived，具体参考 1.1.4 节。

之后配置 Kubernetes 集群中的 Master 组件。

在所有节点创建相关目录：

```
mkdir -p /etc/kubernetes/manifests/ /etc/systemd/system/kubelet.service.d
/var/lib/kubelet /var/log/kubernetes
```

所有 Master 创建 kube-apiserver 文件，主要修改 advertise-address 和 etcd-servers，如果在之前修改了 ClusterIP 的网段（默认 10.96.0.0），此时也要修改 service-cluster-ip-range 的值：

```
[root@K8S-master01 ~]# cat /usr/lib/systemd/system/kube-apiserver.service
[Unit]
Description=Kubernetes API Server
Documentation=https://github.com/kubernetes/kubernetes
After=network.target

[Service]
ExecStart=/usr/local/bin/kube-apiserver \
     --v=2 \
     --logtostderr=true \
     --allow-privileged=true \
     --bind-address=0.0.0.0 \
     --secure-port=6443 \
     --insecure-port=0 \
     --advertise-address=192.168.20.110 \
     --service-cluster-ip-range=10.96.0.0/12 \
     --service-node-port-range=30000-32767 \
     --etcd-servers=https://192.168.20.20:2379,https://192.168.20.21:2379,
https://192.168.20.22:2379 \
     --etcd-cafile=/etc/etcd/ssl/etcd-ca.pem \
     --etcd-certfile=/etc/etcd/ssl/etcd.pem \
     --etcd-keyfile=/etc/etcd/ssl/etcd-key.pem \
     --client-ca-file=/etc/kubernetes/pki/ca.pem \
```

```
            --tls-cert-file=/etc/kubernetes/pki/apiserver.pem \
            --tls-private-key-file=/etc/kubernetes/pki/apiserver-key.pem \
            --kubelet-client-certificate=/etc/kubernetes/pki/apiserver.pem \
            --kubelet-client-key=/etc/kubernetes/pki/apiserver-key.pem \
            --service-account-key-file=/etc/kubernetes/pki/sa.pub \
            --kubelet-preferred-address-types=InternalIP,ExternalIP,Hostname \
            --enable-admission-plugins=Initializers,NamespaceLifecycle,
LimitRanger,ServiceAccount,DefaultStorageClass,DefaultTolerationSeconds,NodeRe
striction,ResourceQuota \
            --authorization-mode=Node,RBAC \
            --enable-bootstrap-token-auth=true \
            --requestheader-client-ca-file=/etc/kubernetes/pki/front-proxy-ca.pem \
            --proxy-client-cert-file=/etc/kubernetes/pki/front-proxy-client.pem \
            --proxy-client-key-file=/etc/kubernetes/pki/front-proxy-
client-key.pem \
            --requestheader-allowed-names=aggregator \
            --requestheader-group-headers=X-Remote-Group \
            --requestheader-extra-headers-prefix=X-Remote-Extra- \
            --requestheader-username-headers=X-Remote-User \
            --token-auth-file=/etc/kubernetes/token.csv

    Restart=on-failure
    RestartSec=10s
    LimitNOFILE=65535

    [Install]
    WantedBy=multi-user.target
```

> **注　意**
>
> Initializers 选项在 kube-apiserver 1.14.x 版本的 --enable-admission-plugins 已停用。

```
[root@K8S-master01 ~]# cat /etc/kubernetes/token.csv
d7d356746b508a1a478e49968fba7947,kubelet-bootstrap,10001,"system:kubelet-bootstrap"
```

所有 Master 节点启动 kube-apiserver：

```
systemctl enable --now kube-apiserver
```

所有 Master 创建 kube-controller-manager.service，注意修改 cluster-cidr 的值，此值为 Pod IP 的网段，不要和宿主机在同一个网段：

```
cat /usr/lib/systemd/system/kube-controller-manager.service
[Unit]
Description=Kubernetes Controller Manager
Documentation=https://github.com/kubernetes/kubernetes
After=network.target

[Service]
ExecStart=/usr/local/bin/kube-controller-manager \
    --v=2 \
    --logtostderr=true \
    --address=127.0.0.1 \
    --root-ca-file=/etc/kubernetes/pki/ca.pem \
```

```
    --cluster-signing-cert-file=/etc/kubernetes/pki/ca.pem \
    --cluster-signing-key-file=/etc/kubernetes/pki/ca-key.pem \
    --service-account-private-key-file=/etc/kubernetes/pki/sa.key \
    --kubeconfig=/etc/kubernetes/controller-manager.kubeconfig \
    --leader-elect=true \
    --use-service-account-credentials=true \
    --node-monitor-grace-period=40s \
    --node-monitor-period=5s \
    --pod-eviction-timeout=2m0s \
    --controllers=*,bootstrapsigner,tokencleaner \
    --allocate-node-cidrs=true \
    --cluster-cidr=10.244.0.0/16 \
    --requestheader-client-ca-file=/etc/kubernetes/pki/front-proxy-ca.pem \
    --node-cidr-mask-size=24

Restart=always
RestartSec=10s

[Install]
WantedBy=multi-user.target
```

所有 Master 节点启动 kube-controller-manager：

```
systemctl daemon-reload
systemctl enable --now kube-controller-manager
```

所有 Master 创建 kube-scheduler：

```
cat /usr/lib/systemd/system/kube-scheduler.service
[Unit]
Description=Kubernetes Scheduler
Documentation=https://github.com/kubernetes/kubernetes
After=network.target

[Service]
ExecStart=/usr/local/bin/kube-scheduler \
    --v=2 \
    --logtostderr=true \
    --address=127.0.0.1 \
    --leader-elect=true \
    --kubeconfig=/etc/kubernetes/scheduler.kubeconfig

Restart=always
RestartSec=10s

[Install]
WantedBy=multi-user.target
```

所有 Master 启动 scheduler：

```
systemctl daemon-reload
systemctl enable --now kube-scheduler
```

查看集群状态。注意如果修改了 ClusterIP 的范围，kubernetes 的端点 Service 就会有所不同，此时采用的是默认的 10.96.0.0 网段：

```
[root@K8S-master01 pki]# cp /etc/kubernetes/admin.kubeconfig ~/.kube/config
[root@K8S-master01 master]# kubectl get cs
NAME                 STATUS    MESSAGE             ERROR
scheduler            Healthy   ok
controller-manager   Healthy   ok
etcd-0               Healthy   {"health":"true"}
etcd-2               Healthy   {"health":"true"}
etcd-1               Healthy   {"health":"true"}
[root@K8S-master01 master]# kubectl get svc
NAME         TYPE        CLUSTER-IP    EXTERNAL-IP   PORT(S)   AGE
kubernetes   ClusterIP   10.96.0.1     <none>        443/TCP   110s
```

此时未配置网络组件，可能会出现如下报错，可忽略：

```
    Jan  2 15:12:55 K8S-master02 kubelet: W0102 15:12:55.223447    11575 cni.go:188]
Unable to update cni config: No networks found in /etc/cni/net.d
    Jan  2 15:12:55 K8S-master02 kubelet: E0102 15:12:55.223591    11575
kubelet.go:2167] Container runtime network not ready: NetworkReady=false
reason:NetworkPluginNotReady message:docker: network plugin is not ready: cni
config uninitialized
```

1.3.5 TLS Bootstrapping 配置

建立 TLS Bootstrapping RBAC 与 Secret，用来解决手动对每台节点单独签署凭证的问题。

建立 bootstrap-kubelet.conf 的 kubernetes config 文件：

```
    kubectl config set-cluster kubernetes
--certificate-authority=/etc/kubernetes/pki/ca.pem     --embed-certs=true
--server=https://192.168.20.110:8443
--kubeconfig=/etc/kubernetes/bootstrap-kubelet.kubeconfig

    kubectl config set-credentials tls-bootstrap-token-user
--token=405d86.c33faabc271aab37
--kubeconfig=/etc/kubernetes/bootstrap-kubelet.kubeconfig

    kubectl config set-context tls-bootstrap-token-user@kubernetes
--cluster=kubernetes     --user=tls-bootstrap-token-user
--kubeconfig=/etc/kubernetes/bootstrap-kubelet.kubeconfig

    kubectl config use-context tls-bootstrap-token-user@kubernetes
--kubeconfig=/etc/kubernetes/bootstrap-kubelet.kubeconfig
```

建立 bootstrap secret，注意 token-id 与 token-secret 的值和上述命令 set-credentials 的 --token 对应：

```
# cat bootstrap-secret.yaml
apiVersion: v1
kind: Secret
metadata:
  name: bootstrap-token-433f5d
  namespace: kube-system
type: bootstrap.kubernetes.io/token
stringData:
  token-id: '405d86'
```

```yaml
  token-secret: c33faabc271aab37
  usage-bootstrap-authentication: "true"
  usage-bootstrap-signing: "true"
  auth-extra-groups: system:bootstrappers:default-node-token
```

```
[root@K8S-master01 ~]# kubectl create -f bootstrap-secret.yaml
secret/bootstrap-token-433f5d created
```

```yaml
# 建立 bootstrap RBAC:
# cat bootstrap-rbac.yaml
apiVersion: rbac.authorization.K8S.io/v1
kind: ClusterRoleBinding
metadata:
  name: kubelet-bootstrap
roleRef:
  apiGroup: rbac.authorization.K8S.io
  kind: ClusterRole
  name: system:node-bootstrapper
subjects:
- apiGroup: rbac.authorization.K8S.io
  kind: Group
  name: system:bootstrappers:default-node-token
---
apiVersion: rbac.authorization.K8S.io/v1
kind: ClusterRoleBinding
metadata:
  name: node-autoapprove-bootstrap
roleRef:
  apiGroup: rbac.authorization.K8S.io
  kind: ClusterRole
  name: system:certificates.K8S.io:certificatesigningrequests:nodeclient
subjects:
- apiGroup: rbac.authorization.K8S.io
  kind: Group
  name: system:bootstrappers:default-node-token
---
apiVersion: rbac.authorization.K8S.io/v1
kind: ClusterRoleBinding
metadata:
  name: node-autoapprove-certificate-rotation
roleRef:
  apiGroup: rbac.authorization.K8S.io
  kind: ClusterRole
  name: system:certificates.K8S.io:certificatesigningrequests:selfnodeclient
subjects:
- apiGroup: rbac.authorization.K8S.io
  kind: Group
  name: system:nodes
---
apiVersion: rbac.authorization.K8S.io/v1
kind: ClusterRole
metadata:
  annotations:
    rbac.authorization.kubernetes.io/autoupdate: "true"
```

```yaml
  labels:
    kubernetes.io/bootstrapping: rbac-defaults
  name: system:kube-apiserver-to-kubelet
rules:
  - apiGroups:
      - ""
    resources:
      - nodes/proxy
      - nodes/stats
      - nodes/log
      - nodes/spec
      - nodes/metrics
    verbs:
      - "*"
---
apiVersion: rbac.authorization.K8S.io/v1
kind: ClusterRoleBinding
metadata:
  name: system:kube-apiserver
  namespace: ""
roleRef:
  apiGroup: rbac.authorization.K8S.io
  kind: ClusterRole
  name: system:kube-apiserver-to-kubelet
subjects:
  - apiGroup: rbac.authorization.K8S.io
    kind: User
    name: kube-apiserver
```

创建 bootstrap：

```
[root@K8S-master01 1.2.1]# kubectl create -f bootstrap-rbac.yaml
```

1.3.6　Node 节点的配置

Node 节点只需要启动 Kubelet 即可，具体配置步骤如下。

将证书复制到 Node 节点：

```
for NODE in K8S-node01 K8S-node02; do
    ssh $NODE mkdir -p /etc/kubernetes/pki /etc/etcd/ssl /etc/etcd/ssl
    for FILE in etcd-ca.pem etcd.pem etcd-key.pem; do
      scp /etc/etcd/ssl/$FILE $NODE:/etc/etcd/ssl/
    done
    for FILE in pki/ca.pem pki/ca-key.pem pki/front-proxy-ca.pem bootstrap-kubelet.kubeconfig; do
      scp /etc/kubernetes/$FILE $NODE:/etc/kubernetes/${FILE}
    done
done
```

配置 10-kubelet.conf 文件，因为 Node 节点采用自动颁发证书的方式，所以此文件需要添加 KUBELET_KUBECONFIG_ARGS 参数，如果已经配置了，就无需再配置。

所有 Node 节点创建相关目录：

```
mkdir -p /var/lib/kubelet /var/log/kubernetes
/etc/systemd/system/kubelet.service.d /etc/kubernetes/manifests/
```

所有 Node 节点配置 Kubelet，如果 Master 节点也需要运行 Pod（在生产环境中不建议，在测试环境中为了节省资源可以运行 Pod），同样需要配置 kubelet，Master 节点和 Node 节点的 kubelet 配置唯一的区别是 Master 节点的--node-labels 为 node-role.kubernetes.io/master=""，Node 节点的为 node-role.kubernetes.io/node=""。因为 Master 节点已经有证书，所以无需再次复制证书，直接创建 kubelet 的配置文件即可。Kubelet service 文件如下：

```
# cat /usr/lib/systemd/system/kubelet.service
[Unit]
Description=Kubernetes Kubelet
Documentation=https://github.com/kubernetes/kubernetes
After=docker.service
Requires=docker.service

[Service]
ExecStart=/usr/local/bin/kubelet

Restart=always
StartLimitInterval=0
RestartSec=10

[Install]
WantedBy=multi-user.target
```

Kubelet Service 参数文件如下：

```
# cat /etc/systemd/system/kubelet.service.d/10-kubelet.conf
[Service]
Environment=" KUBELET_KUBECONFIG_ARGS=--bootstrap-kubeconfig=/etc/kubernetes/bootstrap-kubelet.kubeconfig
--kubeconfig=/etc/kubernetes/kubelet.kubeconfig --allow-privileged=true"
   Environment="KUBELET_SYSTEM_ARGS=--network-plugin=cni
--cni-conf-dir=/etc/cni/net.d --cni-bin-dir=/opt/cni/bin"
   Environment="KUBELET_CONFIG_ARGS=--config=/etc/kubernetes/kubelet-conf.yml"
   Environment="KUBELET_EXTRA_ARGS=--node-labels=node-role.kubernetes.io/node=''
--pod-infra-container-image=registry.cn-hangzhou.aliyuncs.com/google_containers/pause-amd64:3.1"
   ExecStart=
   ExecStart=/usr/local/bin/kubelet $KUBELET_KUBECONFIG_ARGS
$KUBELET_CONFIG_ARGS $KUBELET_SYSTEM_ARGS $KUBELET_EXTRA_ARGS
```

配置 Kubelet 配置文件。注意 clusterDNS 的地址，如果之前修改了 ClusterIP 的网段，需要将 ClusterDNS 的地址改成同网段的地址：

```
# cat /etc/kubernetes/kubelet-conf.yml
apiVersion: kubelet.config.K8S.io/v1beta1
kind: KubeletConfiguration
address: 0.0.0.0
port: 10250
readOnlyPort: 10255
authentication:
```

```yaml
  anonymous:
    enabled: false
  webhook:
    cacheTTL: 2m0s
    enabled: true
  x509:
    clientCAFile: /etc/kubernetes/pki/ca.pem
authorization:
  mode: Webhook
  webhook:
    cacheAuthorizedTTL: 5m0s
    cacheUnauthorizedTTL: 30s
cgroupDriver: cgroupfs
cgroupsPerQOS: true
clusterDNS:
- 10.96.0.10
clusterDomain: cluster.local
containerLogMaxFiles: 5
containerLogMaxSize: 10Mi
contentType: application/vnd.kubernetes.protobuf
cpuCFSQuota: true
cpuManagerPolicy: none
cpuManagerReconcilePeriod: 10s
enableControllerAttachDetach: true
enableDebuggingHandlers: true
enforceNodeAllocatable:
- pods
eventBurst: 10
eventRecordQPS: 5
evictionHard:
  imagefs.available: 15%
  memory.available: 100Mi
  nodefs.available: 10%
  nodefs.inodesFree: 5%
evictionPressureTransitionPeriod: 5m0s
failSwapOn: true
fileCheckFrequency: 20s
hairpinMode: promiscuous-bridge
healthzBindAddress: 127.0.0.1
healthzPort: 10248
httpCheckFrequency: 20s
imageGCHighThresholdPercent: 85
imageGCLowThresholdPercent: 80
imageMinimumGCAge: 2m0s
iptablesDropBit: 15
iptablesMasqueradeBit: 14
kubeAPIBurst: 10
kubeAPIQPS: 5
makeIPTablesUtilChains: true
maxOpenFiles: 1000000
maxPods: 110
nodeStatusUpdateFrequency: 10s
oomScoreAdj: -999
podPidsLimit: -1
registryBurst: 10
```

```
registryPullQPS: 5
resolvConf: /etc/resolv.conf
rotateCertificates: true
runtimeRequestTimeout: 2m0s
serializeImagePulls: true
staticPodPath: /etc/kubernetes/manifests
streamingConnectionIdleTimeout: 4h0m0s
syncFrequency: 1m0s
volumeStatsAggPeriod: 1m0s
```

所有节点启动 kubelet，启动后会自动生成 kubelet.kubeconfig 文件，并且 controller-manager 会自动为 kubelet 颁发证书：

```
systemctl daemon-reload
systemctl enable --now kubelet
```

查看集群状态：

```
[root@K8S-master01 K8S-ha-install]# kubectl get csr
NAME                                                   AGE   REQUESTOR                      CONDITION
csr-2t5rn                                              26m   system:node:K8S-master03       Approved,Issued
csr-8wr62                                              26m   system:node:K8S-master02       Approved,Issued
csr-f6p65                                              13m   system:node:K8S-master01       Approved,Issued
csr-j6gj8                                              26m   system:node:K8S-master01       Approved,Issued
csr-nmlwx                                              13m   system:node:K8S-master03       Approved,Issued
csr-szzbl                                              13m   system:node:K8S-master02       Approved,Issued
node-csr-73a-dLBIRgyLSEEbJ7d2RRMVKr1UeG0dnSAWy9xglvQ   34s   system:bootstrap:cfba97        Approved,Issued
node-csr-Plr5aXhKbqs8KtzV5dile8kuAIKczwaQrp9OoSbUpzQ   33s   system:bootstrap:cfba97        Approved,Issued
[root@K8S-master01 K8S-ha-install]# kubectl get nodes
NAME            STATUS      ROLES    AGE    VERSION
K8S-master01    NotReady    master   26m    v1.13.1
K8S-master02    NotReady    master   26m    v1.13.1
K8S-master03    NotReady    master   26m    v1.13.1
K8S-node01      NotReady    node     40s    v1.13.1
K8S-node02      NotReady    node     42s    v1.13.1
```

1.3.7 Kube-Proxy 配置

Kube-Proxy 用于实现 Pod 和 Pod 之间以及外部到 Pod 的访问，这些访问主要有三种实现方式，在部署过程中无须理会实现方式，具体原理请参考本书的 2.2.13 节。

以下介绍 kube-Proxy 的配置步骤。

创建 Kube-Proxy 的 ServiceAccount：

```
[root@K8S-master01 K8S-ha-install]# kubectl -n kube-system create
```

```
serviceaccount kube-proxy
    serviceaccount/kube-proxy created
```

创建 Clusterrolebinding：
```
kubectl create clusterrolebinding system:kube-proxy \
    --clusterrole system:node-proxier \
    --serviceaccount kube-system:kube-proxy
```

创建 kube-proxy 的 kubeconfig：

```
SECRET=$(kubectl -n kube-system get sa/kube-proxy \
    --output=jsonpath='{.secrets[0].name}')

JWT_TOKEN=$(kubectl -n kube-system get secret/$SECRET \
--output=jsonpath='{.data.token}' | base64 -d)
PKI_DIR=/etc/kubernetes/pki
K8S_DIR=/etc/kubernetes
# proxy set cluster
kubectl config set-cluster kubernetes \
    --certificate-authority=/etc/kubernetes/pki/ca.pem \
    --embed-certs=true \
    --server=https://192.168.20.110:8443 \
    --kubeconfig=${K8S_DIR}/kube-proxy.kubeconfig

# proxy set credentials
kubectl config set-credentials kubernetes \
    --token=${JWT_TOKEN} \
    --kubeconfig=/etc/kubernetes/kube-proxy.kubeconfig

# proxy set context
kubectl config set-context kubernetes \
    --cluster=kubernetes \
    --user=kubernetes \
    --kubeconfig=/etc/kubernetes/kube-proxy.kubeconfig

# proxy set default context
kubectl config use-context kubernetes \
    --kubeconfig=/etc/kubernetes/kube-proxy.kubeconfig
```

复制 kube-proxy 的文件至所有节点：

```
cd chap01/1.3
for NODE in K8S-master01 K8S-master02 K8S-master03; do
    scp ${K8S_DIR}/kube-proxy.kubeconfig $NODE:/etc/Kubernetes/kube-proxy.kubeconfig
    scp kube-proxy/kube-proxy.conf $NODE:/etc/kubernetes/kube-proxy.conf
    scp kube-proxy/kube-proxy.service $NODE:/usr/lib/systemd/system/kube-proxy.service
done

for NODE in K8S-node01 K8S-node02; do
    scp /etc/Kubernetes/kube-proxy.kubeconfig $NODE:/etc/Kubernetes/kube-proxy.kubeconfig
    scp kube-proxy/kube-proxy.conf $NODE:/etc/kubernetes/kube-proxy.conf
    scp kube-proxy/kube-proxy.service $NODE:/usr/lib/systemd/system/kube-proxy.service
```

```
done
```

所有节点启动 kube-proxy：

```
systemctl enable --now kube-proxy
```

1.3.8 Calico 配置

安装 Calico，请参考 1.2.3 节，更改<your-pod-cidr>的值为上述创建的 PodIP 网段：

```
POD_CIDR="<your-pod-cidr>" \
sed -i -e "s?192.168.0.0/16?$POD_CIDR?g" calico/v3.6.1/calico.yaml
```

创建 Calico：

```
[root@K8S-master01 1.2.1]# kubectl create -f calico/v3.6.1/
configmap/calico-config created
service/calico-typha created
deployment.apps/calico-typha created
daemonset.extensions/calico-node created
customresourcedefinition.apiextensions.K8S.io/felixconfigurations.crd.proj
ectcalico.org created
customresourcedefinition.apiextensions.K8S.io/bgppeers.crd.projectcalico.o
rg created
customresourcedefinition.apiextensions.K8S.io/bgpconfigurations.crd.projec
tcalico.org created
customresourcedefinition.apiextensions.K8S.io/ippools.crd.projectcalico.or
g created
customresourcedefinition.apiextensions.K8S.io/hostendpoints.crd.projectcal
ico.org created
customresourcedefinition.apiextensions.K8S.io/clusterinformations.crd.proj
ectcalico.org created
customresourcedefinition.apiextensions.K8S.io/globalnetworkpolicies.crd.pr
ojectcalico.org created
customresourcedefinition.apiextensions.K8S.io/globalnetworksets.crd.projec
tcalico.org created
customresourcedefinition.apiextensions.K8S.io/networkpolicies.crd.projectc
alico.org created
   serviceaccount/calico-node created
   serviceaccount/calicoctl created
   deployment.extensions/calicoctl created
   clusterrole.rbac.authorization.K8S.io/calicoctl created
   clusterrolebinding.rbac.authorization.K8S.io/calicoctl created
   clusterrole.rbac.authorization.K8S.io/calico-node created
   clusterrolebinding.rbac.authorization.K8S.io/calico-node created
```

查看 Calico Pods 的状态：

```
[root@K8S-master01 1.2.1]# kubectl get po -n kube-system
NAME                     READY    STATUS     RESTARTS    AGE
calico-node-47rzn         2/2     Running      0          48s
calico-node-799q4         2/2     Running      0          48s
calico-node-8gmvz         2/2     Running      0          48s
calico-node-mkv8n         2/2     Running      0          48s
calico-node-n9txr         2/2     Running      0          48s
```

```
calicoctl-66d787bc49-cjq4w    1/1     Running    0          49s
```

查看集群的状态，此时报错日志已解决：

```
[root@K8S-master01 1.2.1]# kubectl get no
NAME            STATUS   ROLES    AGE    VERSION
K8S-master01    Ready    master   66m    v1.13.1
K8S-master02    Ready    master   66m    v1.13.1
K8S-master03    Ready    master   66m    v1.13.1
K8S-node01      Ready    node     41m    v1.13.1
K8S-node02      Ready    node     41m    v1.13.1
```

1.3.9 CoreDNS 的配置

CoreDNS 用于集群中 Pod 解析 Service 的名字，Kubernetes 基于 CoreDNS 用于服务发现功能。安装 CoreDNS 1.3.1 版本（写本书时的最新版）。如果之前修改了 PodIP 的网段，需要自行修改此文件的 ClusterIP 参数：

```
[root@K8S-master01 1.2.1]# kubectl apply -f coredns/coredns.yml
serviceaccount/coredns created
clusterrole.rbac.authorization.K8S.io/system:coredns created
clusterrolebinding.rbac.authorization.K8S.io/system:coredns created
configmap/coredns created
deployment.extensions/coredns created
service/kube-dns created
```

查看 CoreDNS 的 Pods 状态：

```
[root@K8S-master01 1.2.1]# kubectl -n kube-system get po -l K8S-app=kube-dns
NAME                        READY   STATUS    RESTARTS   AGE
coredns-647d577766-fnr8j    1/1     Running   0          110s
coredns-647d577766-lw8qj    1/1     Running   0          110s
```

解析测试：

```
cat<<EOF | kubectl apply -f -
apiVersion: v1
kind: Pod
metadata:
  name: busybox
  namespace: default
spec:
  containers:
  - name: busybox
    image: busybox:1.28
    command:
      - sleep
      - "3600"
    imagePullPolicy: IfNotPresent
  restartPolicy: Always
EOF
```

解析 Kubernetes service：

```
[root@K8S-master02 ~]# kubectl exec -ti busybox -n default -- nslookup
```

```
kubernetes
    Server:     10.96.0.10
    Address 1: 10.96.0.10 kube-dns.kube-system.svc.cluster.local

    Name:       kubernetes
    Address 1: 10.96.0.1 kubernetes.default.svc.cluster.local
```

1.3.10 Metrics-Server 配置

安装 metrics-server：

```
[root@K8S-master01 1.2.1]# kubectl create -f metric-server/
clusterrole.rbac.authorization.K8S.io/system:aggregated-metrics-reader created
clusterrolebinding.rbac.authorization.K8S.io/metrics-server:system:auth-delegator created
rolebinding.rbac.authorization.K8S.io/metrics-server-auth-reader created
apiservice.apiregistration.K8S.io/v1beta1.metrics.K8S.io created
serviceaccount/metrics-server created
deployment.extensions/metrics-server created
service/metrics-server created
clusterrole.rbac.authorization.K8S.io/system:metrics-server created
clusterrolebinding.rbac.authorization.K8S.io/system:metrics-server created
```

查看 Pod 状态：

```
[root@K8S-master01 1.2.1]# kubectl -n kube-system get po -l K8S-app=metrics-server
NAME                              READY   STATUS    RESTARTS   AGE
metrics-server-5fbf6d6cb7-k57mf   1/1     Running   0          4m39s
```

查看 Node 资源使用：

```
[root@K8S-master01 1.2.1]# kubectl top node
NAME           CPU(cores)   CPU%   MEMORY(bytes)   MEMORY%
K8S-master01   95m          2%     1123Mi          29%
K8S-master02   128m         3%     1140Mi          29%
K8S-master03   158m         3%     1246Mi          32%
K8S-node01     121m         3%     398Mi           5%
K8S-node02     115m         2%     382Mi           4%
```

1.3.11 Dashboard 配置

安装 heapster：

```
[root@k8s-master01 1.2.1]# kubectl apply -f heapster/
```

安装 Dashboard：

```
[root@K8S-master01 1.2.1]# kubectl create -f dashboard/
clusterrole.rbac.authorization.K8S.io/anonymous-dashboard-proxy-role created
clusterrolebinding.rbac.authorization.K8S.io/anonymous-dashboard-proxy-binding created
```

```
serviceaccount/dashboard created
clusterrolebinding.rbac.authorization.K8S.io/dashboard created
secret/kubernetes-dashboard-certs created
serviceaccount/kubernetes-dashboard created
role.rbac.authorization.K8S.io/kubernetes-dashboard-minimal created
rolebinding.rbac.authorization.K8S.io/kubernetes-dashboard-minimal created
deployment.apps/kubernetes-dashboard created
service/kubernetes-dashboard created
serviceaccount/admin-user created
clusterrolebinding.rbac.authorization.K8S.io/admin-user created
```

查看 Pod 的状态：

```
[root@K8S-master01 1.2.1]# kubectl get po -n kube-system -l
K8S-app=kubernetes-dashboard
  NAME                                       READY   STATUS    RESTARTS   AGE
  kubernetes-dashboard-79cfbd6f7f-nvgkh      1/1     Running   0          <invalid>
```

通过 NodePort 访问 https://VIP:30000。

选择令牌，参考图 1-4。

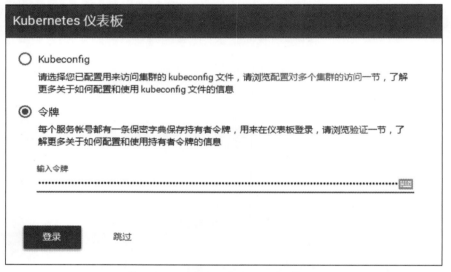

图 1-4　Dashboard 登录方式选择令牌

输入令牌（令牌获取参考 1.1.9 小节）后登录，即可登录到 Dashboard 页面，参考图 1-5。

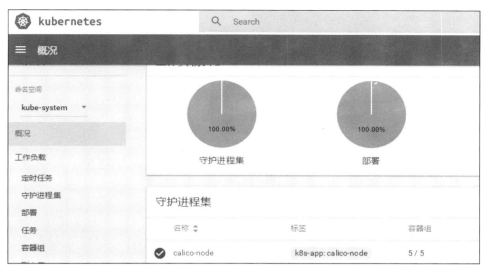

图 1-5　Dashboard 页面

1.4　小　结

本章使用 Kubeadm 和二进制方式安装了高可用的 Kubernetes 集群，至此读者已经打开了 Kubernetes 的第一扇大门，相信读者对 Kubernetes 的理解又深了一些。虽然使用二进制方式安装 Kubernetes 集群过程较为复杂，但这也是每个技术人员必须要掌握的内容，可以让读者深入地了解 Kubernetes 的安装过程。在安装过程中，可以采用自己喜欢的安装方式进行高可用集群的搭建，相信在以后的 Kubernetes 版本中，Kubeadm 安装方式会变得更加方便、稳定。对于新版本的安装，无论是 Kubeadm 还是二进制安装方式，安装步骤大致相同，可能只是一些参数上的变化，读者可以查看对应版本的参数配置进行安装即可。在生产环境中，建议读者使用三台及三台以上奇数个 Master 节点和 Etcd 节点，并且 Etcd 的数据目录（/var/lib/etcd）最好使用固态硬盘，有条件的情况下，对 Docker 的存储目录（/var/lib/docker）也可以单独挂一个高效的存储。对于集群的 Node 节点扩容，只需重复基本配置、基本组件安装和 Node 节点配置即可完成扩容，完成扩容后最好先使用 Taint 禁止调度，对新节点进行各方面的测试后，再加入到集群调度中。

第 2 章

Docker 及 Kubernetes 基础

上一章主要讲解了 Kubernetes 各种版本的安装方式，相信读者已经有了一套高可用 Kubernetes 集群了，并且也对 Kubernetes 的架构和各种组件有了一些认识。本章主要讲解 Docker 和 Kubernetes 的一些基本概念和简单操作，基于上一章搭建的集群来学习本章内容会让自己印象更加深刻。

2.1 Docker 基础

2.1.1 Docker 介绍

Docker 是一个开源的软件项目，在 Linux 操作系统上，Docker 提供了一个额外的软件抽象层及操作系统层虚拟化的自动管理机制。Docker 运行名为"Container（容器）"的软件包，容器之间彼此隔离，并捆绑了自己的应用程序、工具、库和配置文件。所有容器都由单个操作系统内核运行，因此比虚拟机更轻量级。

Docker 利用 Linux 资源分离机制，例如 cgroups 及 Linux Namespace 来创建相互独立的容器（Container），可以在单个 Linux 实体下运行，避免了启动一个虚拟机造成的额外负担。Linux 核心对 Namespace（命名空间）的支持完全隔离了不同 Namespace 下的应用程序的"视野"（即作用范围），包括进程树、网络、用户 ID 与挂载的文件系统等，而核心 cgroups 则提供了资源隔离，包括 CPU、存储器、Block I/O 与网络。

2.1.2 Docker 基本命令

本节介绍 Docker 的一些常用命令，这些命令有助于读者排查和解决集群中的问题。

查看 Docker 版本。包括 Docker 版本号、API 版本号、Git Commit、Go 版本号等。

```
[root@K8S-master01 ~]# docker version
```

```
Client:
 Version:      17.09.1-ce
 API version:  1.32
 Go version:   go1.8.3
 Git commit:   19e2cf6
 Built:        Thu Dec  7 22:23:40 2017
 OS/Arch:      linux/amd64

Server:
 Version:      17.09.1-ce
 API version:  1.32 (minimum version 1.12)
 Go version:   go1.8.3
 Git commit:   19e2cf6
 Built:        Thu Dec  7 22:25:03 2017
 OS/Arch:      linux/amd64
 Experimental: false
```

显示 Docker 信息:

```
Containers: 22
 Running: 21
 Paused: 0
 Stopped: 1
Images: 18
Server Version: 17.09.1-ce
Storage Driver: overlay2
 Backing Filesystem: xfs
 Supports d_type: true
 Native Overlay Diff: true
Logging Driver: json-file
Cgroup Driver: cgroupfs
Plugins:
 Volume: local
 Network: bridge host macvlan null overlay
 Log: awslogs fluentd gcplogs gelf journald json-file logentries splunk syslog
Swarm: inactive
Runtimes: runc
Default Runtime: runc
Init Binary: docker-init
containerd version: 06b9cb35161009dcb7123345749fef02f7cea8e0
runc version: 3f2f8b84a77f73d38244dd690525642a72156c64
init version: 949e6fa
Security Options:
 seccomp
  Profile: default
Kernel Version: 4.18.9-1.el7.elrepo.x86_64
Operating System: CentOS Linux 7 (Core)
OSType: linux
Architecture: x86_64
CPUs: 4
Total Memory: 3.848GiB
Name: K8S-master01
ID: HM66:LH4K:PNES:GFJX:TKNX:TLOH:WONE:KLHT:YRB3:3KAR:3WZJ:HYOX
Docker Root Dir: /var/lib/docker
Debug Mode (client): false
```

```
Debug Mode (server): false
Username: dotbalo
Registry: https://index.docker.io/v1/
Experimental: false
Insecure Registries:
 127.0.0.0/8
Live Restore Enabled: false
```

查询镜像。OFFICIAL 为 OK 的是官方镜像，默认搜索的是 hub.docker.com。

```
[root@K8S-master01 ~]# docker search nginx
  NAME                                              DESCRIPTION
STARS           OFFICIAL        AUTOMATED
  nginx                                             Official build of Nginx.
10749           [OK]
  jwilder/nginx-proxy                               Automated Nginx reverse
proxy for docker c...    1507                       [OK]
  richarvey/nginx-php-fpm                           Container running Nginx +
PHP-FPM capable ...      675                        [OK]
  jrcs/letsencrypt-nginx-proxy-companion            LetsEncrypt container
to use with nginx as...  469                        [OK]
  webdevops/php-nginx                               Nginx with PHP-FPM
120                                     [OK]
  kitematic/hello-world-nginx                       A light-weight nginx
container that demons...  119
  zabbix/zabbix-web-nginx-mysql                     Zabbix frontend based on
Nginx web-server ...    86                          [OK]
  bitnami/nginx                                     Bitnami nginx Docker Image
60                                      [OK]
  linuxserver/nginx                                 An Nginx container, brought
to you by Linu...       51
  1and1internet/ubuntu-16-nginx-php-phpmyadmin-mysql-5
ubuntu-16-nginx-php-phpmyadmin-mysql-5       48
[OK]
  tobi312/rpi-nginx                                 NGINX on Raspberry Pi / armhf
23                                      [OK]
  nginx/nginx-ingress                               NGINX Ingress Controller
for Kubernetes          15
  blacklabelops/nginx                               Dockerized Nginx Reverse
Proxy Server.           12                          [OK]
  wodby/drupal-nginx                                Nginx for Drupal container
image                   11                          [OK]
  centos/nginx-18-centos7                           Platform for running nginx
1.8 or building...      10
  nginxdemos/hello                                  NGINX webserver that serves
a simple page ...        9                          [OK]
  webdevops/nginx                                   Nginx container
8                                       [OK]
  centos/nginx-112-centos7                          Platform for running nginx
1.12 or buildin...       6
  1science/nginx                                    Nginx Docker images that
include Consul Te...     4                          [OK]
  travix/nginx                                      NGinx reverse proxy
2                                       [OK]
  mailu/nginx                                       Mailu nginx frontend
```

```
2                                             [OK]
    pebbletech/nginx-proxy                             nginx-proxy sets up a
container running ng...     2                                 [OK]
    toccoag/openshift-nginx                            Nginx reverse proxy for
Nice running on sa...      1                                  [OK]
    ansibleplaybookbundle/nginx-apb                    An APB to deploy NGINX
0                                             [OK]
    wodby/nginx                                        Generic nginx
```

拉取/下载镜像。默认是 hub.docker.com（docker.io）上面的镜像，如果拉取公司内部的镜像或者其他仓库上的镜像，需要在镜像前面加上仓库的 URL，如：

```
docker pull harbor.xxx.net/frontend:v1
```

拉取公网上的 Nginx 镜像：

```
# 把公网上的镜像拉取到本地服务器，不指定版本号为 latest
[root@K8S-master01 ~]# docker pull nginx
Using default tag: latest
latest: Pulling from library/nginx
Digest: sha256:b543f6d0983fbc25b9874e22f4fe257a567111da96fd1d8f1b44315f1236398c
Status: Image is up to date for nginx:latest

#拉取指定版本
[root@K8S-master01 ~]# docker pull nginx:1.15
1.15: Pulling from library/nginx
Digest: sha256:b543f6d0983fbc25b9874e22f4fe257a567111da96fd1d8f1b44315f1236398c
Status: Downloaded newer image for nginx:1.15
```

推送镜像。把本地的镜像推送到公网仓库中，或者公司内部的仓库中。

默认登录和推送的是公网的镜像，如果需要推送到公司仓库或者其他仓库，只需要在镜像前面使用 tag 并加上 URL 即可：

```
    [root@K8S-master01 ~]# docker images | grep nginx-v2
    nginx-v2                                                        latest
3d9c6e44d3db        3 hours ago         109MB
    [root@K8S-master01 ~]# docker tag nginx-v2 dotbalo/nginx-v2:test
    [root@K8S-master01 ~]# docker images | grep nginx-v2
    dotbalo/nginx-v2                                                test
3d9c6e44d3db        3 hours ago         109MB
    nginx-v2                                                        latest
3d9c6e44d3db        3 hours ago         109MB
    [root@K8S-master01 ~]# docker login
    Login with your Docker ID to push and pull images from Docker Hub. If you don't
have a Docker ID, head over to https://hub.docker.com to create one.
    Username (dotbalo): dotbalo
    Password:
    Login Succeeded
    [root@K8S-master01 ~]# docker push dotbalo/nginx-v2:test
    The push refers to a repository [docker.io/dotbalo/nginx-v2]
    2eaa7b5717a2: Mounted from dotbalo/nginx
    a674e06ede38: Mounted from dotbalo/nginx
    b7efe781401d: Mounted from dotbalo/nginx
```

```
c9c2a3696080: Mounted from dotbalo/nginx
7b4e562e58dc: Mounted from dotbalo/nginx
test: digest: sha256:5d749d2b10150426b510d2c3a05a99cf547c2ca1be382e1dbb2f90b68b6bea96 size: 1362
```

前台启动一个容器：

```
[root@DockerTestServer ~]# docker run -ti nginx bash
root@23bc7ccabb09:/#
```

后台启动：

```
[root@DockerTestServer ~]# docker run -tid nginx bash
1bcf5154d5c3a57d92a6796f526eac2cefd962aaca9cf4098689bfe830bb9e5e
```

端口映射。可以将本机的端口映射到容器的端口，比如将本机的 1111 端口映射到容器的 80 端口：

```
[root@DockerTestServer ~]# docker run -ti -p 1111:80 nginx bash
root@cd676d572188:/#
```

挂载卷。可以将本机的目录挂载到容器的指定目录，比如将 hosts 文件挂载到容器的 hosts：

```
[root@DockerTestServer ~]# docker run -ti -p 1111:80 -v /etc/hosts:/etc/hosts nginx bash
root@cd676d572188:/#
```

查看当前正在运行的容器：

```
[root@K8S-master01 K8S-ha-install]# docker ps
  CONTAINER ID        IMAGE
COMMAND                  CREATED             STATUS              PORTS
NAMES
   862e82066496        94ec7e53edfc
"nginx -g 'daemon ..."   21 hours ago        Up 21 hours
K8S_nginx_nginx-deployment-57895845b8-vb7bs_default_d0d254f8-1fb3-11e9-a9f2-000c293ad492_1
   10bf838e18d0        registry.cn-hangzhou.aliyuncs.com/google_containers/pause-amd64:3.1   "/pause"
21 hours ago        Up 21 hours
K8S_POD_nginx-deployment-57895845b8-vb7bs_default_d0d254f8-1fb3-11e9-a9f2-000c293ad492_1
```

查看所有容器，包括已经退出的：

```
[root@K8S-master01 K8S-ha-install]# docker ps -a
```

查看正在运行的容器（即显示出容器的 ID）：

```
[root@K8S-master01 K8S-ha-install]# docker ps -q
……
0d1a98b3c402
c1fd8ff1f7f2
86b1c069024b
……
```

查看所有容器的 ID，包括已经退出的：

```
[root@K8S-master01 K8S-ha-install]# docker ps -aq
……
17019738d93d
b3bb2a592dfb
e0637b76afe3
0b74e028d0ae
65a1b5e1e501
……
```

进入到一个后台运行的容器（即之前用-d命令参数来指定后台运行方式的容器）：

```
[root@K8S-master01 K8S-ha-install]# docker ps | tail -1
86b1c069024b        nginx:latest
"nginx -g 'daemon ..."   4 days ago          Up 21 hours            80/tcp,
0.0.0.0:16443->16443/tcp   nginx-lb
[root@K8S-master01 K8S-ha-install]# docker exec -ti 86b1c069024b bash
root@nginx-lb:/#
```

拷贝文件。双向拷贝，可以将本机的文件拷贝到容器，反之亦然：

```
[root@K8S-master01 K8S-ha-install]# docker cp README.md
92aceec0dcdd327a709bf0ec83:/tmp
    #exec 也可直接执行容器命令
[root@K8S-master01 K8S-ha-install]# docker exec 92aceec0dcdd327a709bf0ec83 ls
/tmp/
README.md
```

删除已经退出的容器：

```
[root@K8S-master01 K8S-ha-install]# docker ps -a |grep Exited | tail -3
600e5da5c196        3cab8e1b9802
"etcd --advertise-..."   4 days ago          Exited (137) 21 hours ago
K8S_etcd_etcd-K8S-master01_kube-system_c94bb8ceba1b924e6e3175228b168fe0_0
    5a1848d923a1
registry.cn-hangzhou.aliyuncs.com/google_containers/pause-amd64:3.1   "/pause"
4 days ago          Exited (0) 21 hours ago
K8S_POD_kube-scheduler-K8S-master01_kube-system_9c27268d8e3e5c14fa0160192a2c79
88_0
    280fc86494f1
registry.cn-hangzhou.aliyuncs.com/google_containers/pause-amd64:3.1   "/pause"
4 days ago          Exited (0) 21 hours ago
K8S_POD_etcd-K8S-master01_kube-system_c94bb8ceba1b924e6e3175228b168fe0_0
[root@K8S-master01 K8S-ha-install]# docker rm 600e5da5c196 5a1848d923a1
280fc86494f1
    600e5da5c196
    5a1848d923a1
    280fc86494f1
[root@K8S-master01 K8S-ha-install]# docker ps -a |grep Exited | grep -E
"600e5da5c196|5a1848d923a1|280fc86494f1"
```

删除本机镜像。比如删除 REPOSITORY 为 none 的镜像：

```
[root@K8S-master01 K8S-ha-install]# docker images | grep none
    <none><none>                       7ad745acca31        2 days ago          5.83MB
    dotbalo/canary                                                             <none>
00f40cc9b7f6        2 days ago          5.83MB
    dotbalo/canary                                                             <none>
```

```
9b0f2f308931          2 days ago          5.83MB
    <none><none>                      c3d2357e9cbd        2 days ago         4.41MB
    dotbalo/nginx                                                            <none>
97c97cee03f9          3 days ago          109MB
    [root@K8S-master01 K8S-ha-install]# docker rmi 7ad745acca31 00f40cc9b7f6
9b0f2f308931 c3d2357e9cbd 97c97cee03f9
    Deleted:
sha256:7ad745acca31e3f753a3d50e45b7868e9a1aa177369757a9724bccf0654abcb2
    Deleted:
sha256:0546dcf8a97e167875d6563ef7f02ddd8ad3fc0d5f5c064b41e1ce67369b7e06
    Untagged:
dotbalo/canary@sha256:cdd99e578cb2cb8e84eaf2e077c2195a40948c9621d32004a9b5f4e8
2a408f4d
    Deleted:
sha256:00f40cc9b7f6946f17a0eb4fef859aa4e898d3170f023171d0502f8b447353a6
    Deleted:
sha256:7306c50196b5adc635e59152851dbb7fb2dc8782ecb217702849be26e3b1f2a5
    Deleted:
sha256:6b4fe6af6a9cd0d567326e718b91fdd5aca3d39d32bd40bbdd372430be286e3f
    Deleted:
sha256:b864518ff0e99c77046a58f6d82311c8eb64a88ed60bc28d8bd330137eddc024
    Untagged:
dotbalo/canary@sha256:8edea17bdeb346d20f1e93d0d4bf340f42ee8c8373885aa388c536e1
a718c7e7
    Deleted:
sha256:9b0f2f308931a88a5731955d58ae1226b5c147d8f372dae7c2250c0ff9854bf4
    Deleted:
sha256:659bcc00764060582794181890c8b63d6bbf60e8d3da035f76aa2f4d261742d7
    Deleted:
sha256:96163717e76d4b869461e39ed33c4e4066e7de44974557c6206d0f855fb58eb2
    Deleted:
sha256:4a04bebd433278ce549d9e941c2fc3f14021a450ed8ecf58f79c1668b6b9e72e
    Deleted:
sha256:5d65989598faa4ab6361db2655ab43866df88d850621d607474c165eefd6c73e
    Deleted:
sha256:ea8d75cec9b5fc0baf635c584fe818ba9fb2264a30f7210da2c58cfd71cc53b8
    Deleted:
sha256:c3d2357e9cbda84bc7feb1bbebbd3bd9bf6cd37b4415f8746f9cd95b8c11eb83
    Deleted:
sha256:6dc63f7195aae9d4b6764094fe786e32d590851f9646434d29d4fd40acf1c8ef
    Deleted:
sha256:03deec3a1538718aca4021e3f11293c55937fe1191e63a2c59948032e8ded166
    Deleted:
sha256:de87bb7eb02235ddc48979aee72779582ccf07e6b68685d905f8444e3cb5ed94
    Deleted:
sha256:7446f95fc910f657a872f3b99f4b0a91c8aa5471b90aa469a289d3b06f4be22c
    Deleted:
sha256:e901da9c8b00dd031d9ab42623f149a5247b92d78bee375e22ecb67f3b5911c3
    Deleted:
sha256:c9014ca736145dc855ed49b2d11e10fc68b1bcd94b2bf7fa43066d490ae0a7e3
    Deleted:
sha256:f8a9cd62cbd033d5f0cc292698c2ace8f9ca2e322dced364a8b6cbc67dd5d279
    Untagged:
dotbalo/nginx@sha256:deb5bcbfcedf451ddba3422a95b213613dc23c42ee6a63e746d09e040
e0cc7f8
```

```
    Deleted:
sha256:97c97cee03f9a552e4edf34766af09b7f6a74782776a199c5e7492971309158a
    Deleted:
sha256:697f26740b36e9a5aee72a4ca01cc6f644b59092d49ae043de9857e09ca9637e
```

镜像打标签（tag）。用于区分不同版本的镜像：

```
[root@K8S-master01 K8S-ha-install]# docker images | grep nginx | tail -1
nginx
1.7.9                   84581e99d807        3 years ago         91.7MB
#不加 URL 一般为公网仓库中自己的仓库
[root@K8S-master01 K8S-ha-install]# docker tag nginx dotbalo/nginx:v1
#加 URL 一般为公司内部仓库或者其他仓库
[root@K8S-master01 K8S-ha-install]# docker tag nginx
harbor.xxx.net/stage/nginx:v1
```

使用 dockerbuild 通过 Dockerfile 制作镜像。注意最后的一个点（.），表示使用当前目录的 Dockerfile：

```
dockerbuild-t image_name:image_tag .
```

上述演示的都是 Docker 常用的基本命令，已可以满足日常需求，如果读者想要深入了解，可以参考 Docker 的相关资料。

2.1.3 Dockerfile 的编写

Dockerfile 是用来快速创建自定义镜像的一种文本格式的配置文件，在持续集成和持续部署时，需要使用 Dockerfile 生成相关应用程序的镜像，然后推送到公司内部仓库中，再通过部署策略把镜像部署到 Kubernetes 中。

通过 Dockerfile 提供的命令可以构建 Dockerfile 文件，Dockerfile 的常用命令如下：

```
FROM: 继承基础镜像
MAINTAINER: 镜像制作作者的信息
RUN: 用来执行 shell 命令
EXPOSE: 暴露端口号
CMD: 启动容器默认执行的命令，会被覆盖
ENTRYPOINT: 启动容器真正执行的命令，不会被覆盖
VOLUME: 创建挂载点
ENV: 配置环境变量
ADD: 复制文件到容器，一般拷贝文件，压缩包自动解压
COPY: 复制文件到容器，一般拷贝目录
WORKDIR: 设置容器的工作目录
USER: 容器使用的用户
```

以下简单演示每个命令的使用方法。

使用 RUN 创建一个用户：

```
[root@DockerTestServer test]# cat Dockerfile
# base image
FROM centos:6
MAINTAINER dot
RUN useradd dot
```

执行构建

```
docker build -t centos:user .
```

使用 ENV 定义环境变量并用 CMD 执行命令：

```
[root@DockerTestServer test]# cat Dockerfile
# base image
FROM centos:6
MAINTAINER dot
RUN useradd dot
RUN mkdir dot
ENV envir=test version=1.0
CMD echo "envir:$envir version:$version"
```

执行构建并启动测试：

```
#执行构建
docker build -t centos:env-cmd .
#启动镜像验证 ENV 和 CMD
[root@DockerTestServer test]# docker run centos:env-cmd
envir:test version:1.0
```

使用 ADD 添加一个压缩包，使用 WORKDIR 改变工作目录：

```
# base image
FROM nginx
MAINTAINER dot
ADD ./index.tar.gz /usr/share/nginx/html/
WORKDIR /usr/share/nginx/html
```

使用 COPY 拷贝指定目录下的所有文件到容器，不包括本级目录。
此时只会拷贝 webroot 下的所有文件，不会将 webroot 拷贝过去：

```
# base image
FROM nginx
MAINTAINER dot
ADD ./index.tar.gz /usr/share/nginx/html/
WORKDIR /usr/share/nginx/html
COPY webroot/ .
```

设置启动容器的用户，在生产环境中一般不建议使用 root 启动容器，所以可以根据公司业务场景自定义启动容器的用户：

```
# base image
FROM centos:6
MAINTAINER dot

ADD ./index.tar.gz /usr/share/nginx/html/
WORKDIR /usr/share/nginx/html
COPY webroot/ .
RUN useradd -m tomcat -u 1001
USER 1001
```

使用 Volume 创建容器可挂载点：

```
# base image
FROM centos:6
```

```
MAINTAINER dot

VOLUME /data
```

挂载 Web 目录到/data，注意，对于宿主机路径，要写绝对路径：

```
docker run -ti --rm -v `pwd`/web:/data centos:volume bash
```

2.2　Kubernetes 基础

Kubernetes 致力于提供跨主机集群的自动部署、扩展、高可用以及运行应用程序容器的平台，其遵循主从式架构设计，其组件可以分为管理单个节点（Node）组件和控制平面组件。Kubernetes Master 是集群的主要控制单元，用于管理其工作负载并指导整个系统的通信。Kubernetes 控制平面由各自的进程组成，每个组件都可以在单个主节点上运行，也可以在支持高可用集群的多个节点上运行。本节主要介绍 Kubernetes 的重要概念和相关组件。

2.2.1　Master 节点

Master 节点是 Kubernetes 集群的控制节点，在生产环境中不建议部署集群核心组件外的任何 Pod，公司业务的 Pod 更是不建议部署到 Master 节点上，以免升级或者维护时对业务造成影响。
Master 节点的组件包括：

- APIServer。APIServer 是整个集群的控制中枢，提供集群中各个模块之间的数据交换，并将集群状态和信息存储到分布式键-值（key-value）存储系统 Etcd 集群中。同时它也是集群管理、资源配额、提供完备的集群安全机制的入口，为集群各类资源对象提供增删改查以及 watch 的 REST API 接口。APIServer 作为 Kubernetes 的关键组件，使用 Kubernetes API 和 JSON over HTTP 提供 Kubernetes 的内部和外部接口。
- Scheduler。Scheduler 是集群 Pod 的调度中心，主要是通过调度算法将 Pod 分配到最佳的节点（Node），它通过 APIServer 监听所有 Pod 的状态，一旦发现新的未被调度到任何 Node 节点的 Pod（PodSpec.NodeName 为空），就会根据一系列策略选择最佳节点进行调度，对每一个 Pod 创建一个绑定（binding），然后被调度的节点上的 Kubelet 负责启动该 Pod。Scheduler 是集群可插拔式组件，它跟踪每个节点上的资源利用率以确保工作负载不会超过可用资源。因此 Scheduler 必须知道资源需求、资源可用性以及其他约束和策略，例如服务质量、亲和力/反关联性要求、数据位置等。Scheduler 将资源供应与工作负载需求相匹配以维持系统的稳定和可靠，因此 Scheduler 在调度的过程中需要考虑公平、资源高效利用、效率等方面的问题。
- Controller Manager。Controller Manager 是集群状态管理器（它的英文直译名为控制器管理器），以保证 Pod 或其他资源达到期望值。当集群中某个 Pod 的副本数或其他资源因故障和错误导致无法正常运行，没有达到设定的值时，Controller Manager 会尝试自动修复并使其达到期望状态。Controller Manager 包含 NodeController、ReplicationController、

EndpointController、NamespaceController、ServiceAccountController、ResourceQuotaController、ServiceController 和 TokenController，该控制器管理器可与 API 服务器进行通信以在需要时创建、更新或删除它所管理的资源，如 Pod、服务断点等。
- Etcd。Etcd 由 CoreOS 开发，用于可靠地存储集群的配置数据，是一种持久性、轻量型、分布式的键-值（key-value）数据存储组件。Etcd 作为 Kubernetes 集群的持久化存储系统，集群的灾难恢复和状态信息存储都与其密不可分，所以在 Kubernetes 高可用集群中，Etcd 的高可用是至关重要的一部分，在生产环境中建议部署为大于 3 的奇数个数的 Etcd，以保证数据的安全性和可恢复性。Etcd 可与 Master 组件部署在同一个节点上，大规模集群环境下建议部署在集群外，并且使用高性能服务器来提高 Etcd 的性能和降低 Etcd 同步数据的延迟。

2.2.2 Node 节点

Node 节点也被称为 Worker 或 Minion，是主要负责部署容器（工作负载）的单机（或虚拟机），集群中的每个节点都必须具备容器的运行环境（runtime），比如 Docker 及其他组件等。

Kubelet 作为守护进程运行在 Node 节点上，负责监听该节点上所有的 Pod，同时负责上报该节点上所有 Pod 的运行状态，确保节点上的所有容器都能正常运行。当 Node 节点宕机（NotReady 状态）时，该节点上运行的 Pod 会被自动地转移到其他节点上。

Node 节点包括：
- Kubelet，负责与 Master 通信协作，管理该节点上的 Pod。
- Kube-Proxy，负责各 Pod 之间的通信和负载均衡。
- Docker Engine，Docker 引擎，负载对容器的管理。

2.2.3 Pod

1. 什么是 Pod

Pod 可简单地理解为是一组、一个或多个容器，具有共享存储/网络及如何运行容器的规范。Pad 包含一个或多个相对紧密耦合的应用程序容器，处于同一个 Pod 中的容器共享同样的存储空间（Volume，卷或存储卷）、IP 地址和 Port 端口，容器之间使用 localhost:port 相互访问。根据 Docker 的构造，Pod 可被建模为一组具有共享命令空间、卷、IP 地址和 Port 端口的 Docker 容器。

Pod 包含的容器最好是一个容器只运行一个进程。每个 Pod 包含一个 pause 容器，pause 容器是 Pod 的父容器，它主要负责僵尸进程的回收管理。

Kubernetes 为每个 Pod 都分配一个唯一的 IP 地址，这样就可以保证应用程序使用同一端口，避免了发生冲突的问题。

一个 Pod 的状态信息保存在 PodStatus 对象中，在 PodStatus 中有一个 Phase 字段，用于描述 Pod 在其生命周期中的不同状态，参考表 2-1。

表 2-1　Pod 状态字段 Phase 的不同取值

状态	说明
Pending（挂起）	Pod 已被 Kubernetes 系统接收，但仍有一个或多个容器未被创建。可以通过 describe 查看处于 Pending 状态的原因
Running（运行中）	Pod 已经被绑定到一个节点上，并且所有的容器都已经被创建。而且至少有一个是运行状态，或者是正在启动或者重启。可以通过 logs 查看 Pod 的日志
Succeeded（成功）	所有容器执行成功并终止，并且不会再次重启
Failed（失败）	所有容器都已终止，并且至少有一个容器以失败的方式终止，也就是说这个容器要么以非零状态退出，要么被系统终止
Unknown（未知）	通常是由于通信问题造成的无法获得 Pod 的状态

2. Pod 探针

Pod 探针用来检测容器内的应用是否正常，目前有三种实现方式，参考表 2-2。

表 2-2　Pod 探针的实现方式

实现方式	说明
ExecAction	在容器内执行一个指定的命令，如果命令返回值为 0，则认为容器健康
TCPSocketAction	通过 TCP 连接检查容器指定的端口，如果端口开放，则认为容器健康
HTTPGetAction	对指定的 URL 进行 Get 请求，如果状态码在 200~400 之间，则认为容器健康

Pod 探针每次检查容器后可能得到的容器状态，如表 2-3 所示。

表 2-3　Pod 探针检查容器后可能得到的状态

状态	说明
Success（成功）	容器通过检测
Failure（失败）	容器检测失败
Unknown（未知）	诊断失败，因此不采取任何措施

Kubelet 有两种探针（即探测器）可以选择性地对容器进行检测，参考表 2-4。

表 2-4　探针的种类

种类	说明
livenessProbe	用于探测容器是否在运行，如果探测失败，kubelet 会"杀死"容器并根据重启策略进行相应的处理。如果未指定该探针，将默认为 Success
readinessProbe	一般用于探测容器内的程序是否健康，即判断容器是否为就绪（Ready）状态。如果是，则可以处理请求，反之 Endpoints Controller 将从所有的 Services 的 Endpoints 中删除此容器所在 Pod 的 IP 地址。如果未指定，将默认为 Success

3. Pod 镜像拉取策略和重启策略

Pod 镜像拉取策略。用于配置当节点部署 Pod 时，对镜像的操作方式，参考表 2-5。

表 2-5　镜像拉取策略

操作方式	说明
Always	总是拉取，当镜像 tag 为 latest 时，默认为 Always

(续表)

操作方式	说明
Never	不管是否存在都不会拉取
IfNotPresent	镜像不存在时拉取镜像，默认，排除 latest

Pod 重启策略。在 Pod 发生故障时对 Pod 的处理方式参考表 2-6。

表 2-6　Pod 重启策略

操作方式	说明
Always	默认策略。容器失效时，自动重启该容器
OnFailure	容器以不为 0 的状态码终止，自动重启该容器
Never	无论何种状态，都不会重启

4. 创建一个 Pod

在生产环境中，很少会单独启动一个 Pod 直接使用，经常会用 Deployment、DaemonSet、StatefulSet 等方式调度并管理 Pod，定义 Pod 的参数同时适应于 Deployment、DaemonSet、StatefulSet 等方式。

在 Kubeadm 安装方式下，kubernetes 系统组件都是用单独的 Pod 启动的，当然有时候也会单独启动一个 Pod 用于测试业务等，此时可以单独创建一个 Pod。

创建一个 Pod 的标准格式如下：

```yaml
apiVersion: v1　# 必选，API 的版本号
kind: Pod　　　 # 必选，类型 Pod
metadata:　　　 # 必选，元数据
  name: nginx　 # 必选，符合 RFC 1035 规范的 Pod 名称
  namespace: web-testing # 可选，不指定默认为 default，Pod 所在的命名空间
  labels:     # 可选，标签选择器，一般用于 Selector
   - app: nginx
  annotations:  # 可选，注释列表
   - app: nginx
spec:   # 必选，用于定义容器的详细信息
  containers:   # 必选，容器列表
  - name: nginx # 必选，符合 RFC 1035 规范的容器名称
    image: nginx: v1# 必选，容器所用的镜像的地址
    imagePullPolicy: Always # 可选，镜像拉取策略
    command:
    - nginx # 可选，容器启动执行的命令
    - -g
    - "daemon off;"
      workingDir: /usr/share/nginx/html　# 可选，容器的工作目录
      volumeMounts:    # 可选，存储卷配置
      - name: webroot # 存储卷名称
        mountPath: /usr/share/nginx/html # 挂载目录
        readOnly: true    # 只读
      ports:　# 可选，容器需要暴露的端口号列表
      - name: http    # 端口名称
        containerPort: 80 # 端口号
        protocol: TCP # 端口协议，默认 TCP
```

```yaml
    env:         # 可选，环境变量配置
    - name: TZ   # 变量名
      value: Asia/Shanghai
    - name: LANG
      value: en_US.utf8
    resources:   # 可选，资源限制和资源请求限制
      limits:    # 最大限制设置
        cpu: 1000m
        memory: 1024MiB
      requests:  # 启动所需的资源
        cpu: 100m
        memory: 512MiB
    readinessProbe: # 可选，容器状态检查
      httpGet:      # 检测方式
        path: /     # 检查路径
        port: 80    # 监控端口
      timeoutSeconds: 2 # 超时时间
      initialDelaySeconds: 60   # 初始化时间
    livenessProbe:  # 可选，监控状态检查
      exec:  # 检测方式
        command:
        - cat
        - /health
      httpGet:  # 检测方式
        path: /_health
        port: 8080
        httpHeaders:
        - name: end-user
          value: jason
      tcpSocket:    # 检测方式
        port: 80
      initialDelaySeconds: 60   # 初始化时间
      timeoutSeconds: 2 # 超时时间
      periodSeconds: 5  # 检测间隔
      successThreshold: 2 # 检查成功为 2 次表示就绪
      failureThreshold: 1 # 检测失败 1 次表示未就绪
    securityContext:    # 可选，限制容器不可信的行为
      provoleged: false
  restartPolicy: Always # 可选，默认为 Always
  nodeSelector: # 可选，指定 Node 节点
    region: subnet7
  imagePullSecrets: # 可选，拉取镜像使用的 secret
  - name: default-dockercfg-86258
  hostNetwork: false    # 可选，是否为主机模式，如是，会占用主机端口
  volumes:  # 共享存储卷列表
  - name: webroot  # 名称，与上述对应
    emptyDir: {}    # 共享卷类型，空
    hostPath:       # 共享卷类型，本机目录
      path: /etc/hosts
    secret: # 共享卷类型，secret 模式，一般用于密码
      secretName: default-token-tf2jp # 名称
      defaultMode: 420 # 权限
    configMap:    # 一般用于配置文件
```

```
          name: nginx-conf
          defaultMode: 420
```

2.2.4 Label 和 Selector

当 Kubernetes 对系统的任何 API 对象如 Pod 和节点进行"分组"时，会对其添加 Label（key=value 形式的"键-值对"）用以精准地选择对应的 API 对象。而 Selector（标签选择器）则是针对匹配对象的查询方法。注：键-值对就是 key-value pair。

例如，常用的标签 tier 可用于区分容器的属性，如 frontend、backend；或者一个 release_track 用于区分容器的环境，如 canary、production 等。

1. 定义 Label

应用案例：

公司与 xx 银行有一条专属的高速光纤通道，此通道只能与 192.168.7.0 网段进行通信，因此只能将与 xx 银行通信的应用部署到 192.168.7.0 网段所在的节点上，此时可以对节点进行 Label（即加标签）：

```
[root@K8S-master01 ~]# kubectl label node K8S-node02 region=subnet7
node/K8S-node02 labeled
```

然后，可以通过 Selector 对其筛选：

```
[root@K8S-master01 ~]# kubectl get no -l region=subnet7
NAME         STATUS   ROLES    AGE     VERSION
K8S-node02   Ready    <none>   3d17h   v1.12.3
```

最后，在 Deployment 或其他控制器中指定将 Pod 部署到该节点：

```
containers:
  ......
dnsPolicy: ClusterFirst
nodeSelector:
  region: subnet7
restartPolicy: Always
......
```

也可以用同样的方式对 Service 进行 Label：

```
[root@K8S-master01 ~]# kubectl label svc canary-v1 -n canary-production env=canary version=v1
    service/canary-v1 labeled
```

查看 Labels：

```
[root@K8S-master01 ~]# kubectl get svc -n canary-production --show-labels
NAME        TYPE        CLUSTER-IP      EXTERNAL-IP   PORT(S)    AGE   LABELS
canary-v1   ClusterIP   10.110.253.62   <none>        8080/TCP   24h   env=canary,version=v1
```

还可以查看所有 Version 为 v1 的 svc：

```
[root@K8S-master01 canary]# kubectl get svc --all-namespaces -l version=v1
```

```
NAMESPACE              NAME          TYPE        CLUSTER-IP       EXTERNAL-IP   PORT(S)
AGE
canary-production      canary-v1     ClusterIP   10.110.253.62    <none>
8080/TCP    25h
```

其他资源的 Label 方式相同。

2. Selector 条件匹配

Selector 主要用于资源的匹配，只有符合条件的资源才会被调用或使用，可以使用该方式对集群中的各类资源进行分配。

假如对 Selector 进行条件匹配，目前已有的 Label 如下：

```
[root@K8S-master01 ~]# kubectl get svc --show-labels
    NAME          TYPE        CLUSTER-IP        EXTERNAL-IP     PORT(S)      AGE
LABELS
    details       ClusterIP   10.99.9.178       <none>          9080/TCP     45h
app=details
    kubernetes    ClusterIP   10.96.0.1         <none>          443/TCP      3d19h
component=apiserver,provider=kubernetes
    nginx         ClusterIP   10.106.194.137    <none>          80/TCP       2d21h
app=productpage,version=v1
    nginx-v2      ClusterIP   10.108.176.132    <none>          80/TCP       2d20h
<none>
    productpage   ClusterIP   10.105.229.52     <none>          9080/TCP     45h
app=productpage,tier=frontend
    ratings       ClusterIP   10.96.104.95      <none>          9080/TCP     45h
app=ratings
    reviews       ClusterIP   10.102.188.143    <none>          9080/TCP     45h
app=reviews
```

选择 app 为 reviews 或者 productpage 的 svc：

```
[root@K8S-master01 ~]# kubectl get svc -l 'app in (details, productpage)'
--show-labels
    NAME          TYPE        CLUSTER-IP        EXTERNAL-IP     PORT(S)      AGE
LABELS
    details       ClusterIP   10.99.9.178       <none>          9080/TCP     45h
app=details
    nginx         ClusterIP   10.106.194.137    <none>          80/TCP       2d21h
app=productpage,version=v1
    productpage   ClusterIP   10.105.229.52     <none>          9080/TCP     45h
app=productpage,tier=frontend
```

选择 app 为 productpage 或 reviews 但不包括 version=v1 的 svc：

```
[root@K8S-master01 ~]# kubectl get svc -l version!=v1,'app in (details,
productpage)' --show-labels
    NAME          TYPE        CLUSTER-IP        EXTERNAL-IP     PORT(S)      AGE     LABELS
    details       ClusterIP   10.99.9.178       <none>          9080/TCP     45h
app=details
    productpage   ClusterIP   10.105.229.52     <none>          9080/TCP     45h
app=productpage,tier=frontend
```

选择 labelkey 名为 app 的 svc：

```
[root@K8S-master01 ~]# kubectl get svc -l app --show-labels
NAME          TYPE        CLUSTER-IP       EXTERNAL-IP   PORT(S)    AGE    LABELS
details       ClusterIP   10.99.9.178      <none>        9080/TCP   45h    app=details
nginx         ClusterIP   10.106.194.137   <none>        80/TCP     2d21h  app=productpage,version=v1
productpage   ClusterIP   10.105.229.52    <none>        9080/TCP   45h    app=productpage,tier=frontend
ratings       ClusterIP   10.96.104.95     <none>        9080/TCP   45h    app=ratings
reviews       ClusterIP   10.102.188.143   <none>        9080/TCP   45h    app=reviews
```

3. 修改标签（Label）

在实际使用中，Label 的更改是经常发生的事情，可以使用 overwrite 参数修改标签。

修改标签，比如将 version=v1 改为 version=v2：

```
[root@K8S-master01 canary]# kubectl get svc -n canary-production --show-labels
NAME        TYPE        CLUSTER-IP      EXTERNAL-IP   PORT(S)    AGE   LABELS
canary-v1   ClusterIP   10.110.253.62   <none>        8080/TCP   26h   env=canary,version=v1
[root@K8S-master01 canary]# kubectl label svc canary-v1 -n canary-production version=v2 --overwrite
service/canary-v1 labeled
[root@K8S-master01 canary]# kubectl get svc -n canary-production --show-labels
NAME        TYPE        CLUSTER-IP      EXTERNAL-IP   PORT(S)    AGE   LABELS
canary-v1   ClusterIP   10.110.253.62   <none>        8080/TCP   26h   env=canary,version=v2
```

4. 删除标签（Label）

删除标签，比如删除 version：

```
[root@K8S-master01 canary]# kubectl label svc canary-v1 -n canary-production version-
service/canary-v1 labeled
[root@K8S-master01 canary]# kubectl get svc -n canary-production --show-labels
NAME        TYPE        CLUSTER-IP      EXTERNAL-IP   PORT(S)    AGE   LABELS
canary-v1   ClusterIP   10.110.253.62   <none>        8080/TCP   26h   env=canary
```

2.2.5 Replication Controller 和 ReplicaSet

Replication Controller（复制控制器，RC）和 ReplicaSet（复制集，RS）是两种部署 Pod 的方式。因为在生产环境中，主要使用更高级的 Deployment 等方式进行 Pod 的管理和部署，所以本节只对 Replication Controller 和 Replica Set 的部署方式进行简单介绍。

1. Replication Controller

Replication Controller 可确保 Pod 副本数达到期望值，也就是 RC 定义的数量。换句话说，Replication Controller 可确保一个 Pod 或一组同类 Pod 总是可用。

如果存在的 Pod 大于设定的值，则 Replication Controller 将终止额外的 Pod。如果太小，Replication Controller 将启动更多的 Pod 用于保证达到期望值。与手动创建 Pod 不同的是，用 Replication Controller 维护的 Pod 在失败、删除或终止时会自动替换。因此即使应用程序只需要一个 Pod，也应该使用 Replication Controller。Replication Controller 类似于进程管理程序，但是 Replication Controller 不是监视单个节点上的各个进程，而是监视多个节点上的多个 Pod。

定义一个 Replication Controller 的示例如下。

```yaml
apiVersion: v1
kind: ReplicationController
metadata:
  name: nginx
spec:
  replicas: 3
  selector:
    app: nginx
  template:
    metadata:
      name: nginx
      labels:
        app: nginx
    spec:
      containers:
      - name: nginx
        image: nginx
        ports:
        - containerPort: 80
```

2. ReplicaSet

ReplicaSet 是支持基于集合的标签选择器的下一代 Replication Controller，它主要用作 Deployment 协调创建、删除和更新 Pod，和 Replication Controller 唯一的区别是，ReplicaSet 支持标签选择器。在实际应用中，虽然 ReplicaSet 可以单独使用，但是一般建议使用 Deployment（部署）来自动管理 ReplicaSet，除非自定义的 Pod 不需要更新或有其他编排等。

定义一个 ReplicaSet 的示例如下：

```yaml
apiVersion: apps/v1
kind: ReplicaSet
metadata:
  name: frontend
  labels:
    app: guestbook
    tier: frontend
spec:
  # modify replicas according to your case
  replicas: 3
  selector:
    matchLabels:
      tier: frontend
    matchExpressions:
      - {key: tier, operator: In, values: [frontend]}
  template:
    metadata:
```

```yaml
      labels:
        app: guestbook
        tier: frontend
    spec:
      containers:
      - name: php-redis
        image: gcr.io/google_samples/gb-frontend:v3
        resources:
          requests:
            cpu: 100m
            memory: 100Mi
        env:
        - name: GET_HOSTS_FROM
          value: dns
          # If your cluster config does not include a dns service, then to
          # instead access environment variables to find service host
          # info, comment out the 'value: dns' line above, and uncomment the
          # line below.
          # value: env
        ports:
        - containerPort: 80
```

2.2.6　Deployment

虽然 ReplicaSet 可以确保在任何给定时间运行的 Pod 副本达到指定的数量，但是 Deployment（部署）是一个更高级的概念，它管理 ReplicaSet 并为 Pod 和 ReplicaSet 提供声明性更新以及许多其他有用的功能，所以建议在实际使用中，使用 Deployment 代替 ReplicaSet。

如果在 Deployment 对象中描述了所需的状态，Deployment 控制器就会以可控制的速率将实际状态更改为期望状态。也可以在 Deployment 中创建新的 ReplicaSet，或者删除现有的 Deployment 并使用新的 Deployment 部署所用的资源。

1. 创建 Deployment

创建一个 Deployment 文件，并命名为 dc-nginx.yaml，用于部署三个 Nginx Pod：

```yaml
apiVersion: apps/v1
kind: Deployment
metadata:
  name: nginx-deployment
  labels:
    app: nginx
spec:
  replicas: 3
  selector:
    matchLabels:
      app: nginx
  template:
    metadata:
      labels:
        app: nginx
    spec:
      containers:
```

```
      - name: nginx
        image: nginx:1.7.9
        ports:
        - containerPort: 80
```

示例解析

- nginx-deployment：Deployment 的名称。
- replicas：创建 Pod 的副本数。
- selector：定义 Deployment 如何找到要管理的 Pod，与 template 的 label（标签）对应。
- template 字段包含以下字段：
  - app nginx 使用 label（标签）标记 Pod。
  - spec 表示 Pod 运行一个名字为 nginx 的容器。
  - image 运行此 Pod 使用的镜像。
  - Port 容器用于发送和接收流量的端口。

使用 kubectlcreate 创建此 Deployment：

```
[root@K8S-master01 2.2.8.1]# kubectl create -f dc-nginx.yaml
deployment.apps/nginx-deployment created
```

使用 kubectlget 或者 kubectldescribe 查看此 Deployment：

```
[root@K8S-master01 2.2.8.1]# kubectl get deploy
NAME               DESIRED   CURRENT   UP-TO-DATE   AVAILABLE   AGE
nginx-deployment   3         3         3            1           60s
```

其中，

- NAME：集群中 Deployment 的名称。
- DESIRED：应用程序副本数。
- CURRENT：当前正在运行的副本数。
- UP-TO-DATE：显示已达到期望状态的被更新的副本数。
- AVAILABLE：显示用户可以使用的应用程序副本数，当前为 1，因为部分 Pod 仍在创建过程中。
- AGE：显示应用程序运行的时间。

查看此时 Deployment rollout 的状态：

```
[root@K8S-master01 2.2.8.1]# kubectl rollout status
deployment/nginx-deployment
deployment "nginx-deployment" successfully rolled out
```

再次查看此 Deployment：

```
[root@K8S-master01 2.2.8.1]# kubectl get deploy
NAME               DESIRED   CURRENT   UP-TO-DATE   AVAILABLE   AGE
nginx-deployment   3         3         3            3           11m
```

查看此 Deployment 创建的 ReplicaSet：

```
[root@K8S-master01 2.2.8.1]# kubectl get rs
```

```
NAME                              DESIRED    CURRENT    READY    AGE
nginx-deployment-5c689d88bb       3          3          3        12m
```

> **注　意**
>
> ReplicaSet（复制集，RS）的命名格式为[DEPLOYMENT-NAME]-[POD-TEMPLATE-HASH-VALUE]POD- TEMPLATE-HASH-VALUE，是自动生成的，不要手动指定。

查看此 Deployment 创建的 Pod：

```
[root@K8S-master01 2.2.8.1]# kubectl get pods --show-labels
NAME                                    READY   STATUS    RESTARTS   AGE   LABELS
nginx-deployment-5c689d88bb-6b95k       1/1     Running   0          13m   app=nginx,pod-template-hash=5c689d88bb
nginx-deployment-5c689d88bb-9z5z2       1/1     Running   0          13m   app=nginx,pod-template-hash=5c689d88bb
nginx-deployment-5c689d88bb-jc8hr       1/1     Running   0          13m   app=nginx,pod-template-hash=5c689d88bb
```

2. 更新 Deployment

一般对应用程序升级或者版本迭代时，会通过 Deployment 对 Pod 进行滚动更新。

> **注　意**
>
> 当且仅当 Deployment 的 Pod 模板（即.spec.template）更改时，才会触发 Deployment 更新，例如更新 label（标签）或者容器的 image（镜像）。

假如更新 Nginx Pod 的 image 使用 nginx:1.9.1：

```
[root@K8S-master01 2.2.8.1]# kubectl set image deployment nginx-deployment nginx=nginx:1.9.1 --record
deployment.extensions/nginx-deployment image updated
```

当然也可以直接编辑 Deployment，效果相同：

```
[root@K8S-master01 2.2.8.1]# kubectl edit deployment.v1.apps/nginx-deployment
deployment.apps/nginx-deployment edited
```

使用 kubectl rollout status 查看更新状态：

```
[root@K8S-master01 2.2.8.1]# kubectl rollout status deployment.v1.apps/nginx-deployment
Waiting for deployment "nginx-deployment" rollout to finish: 1 out of 3 new replicas have been updated...
Waiting for deployment "nginx-deployment" rollout to finish: 2 out of 3 new replicas have been updated...
Waiting for deployment "nginx-deployment" rollout to finish: 2 out of 3 new replicas have been updated...
Waiting for deployment "nginx-deployment" rollout to finish: 2 out of 3 new replicas have been updated...
Waiting for deployment "nginx-deployment" rollout to finish: 1 old replicas are pending termination...
Waiting for deployment "nginx-deployment" rollout to finish: 1 old replicas
```

```
are pending termination...
deployment "nginx-deployment" successfully rolled out
```

查看 ReplicaSet：

```
[root@K8S-master01 2.2.8.1]# kubectl get rs
NAME                          DESIRED   CURRENT   READY   AGE
nginx-deployment-5c689d88bb   0         0         0       34m
nginx-deployment-6987cdb55b   3         3         3       5m14s
```

通过 describe 查看 Deployment 的详细信息：

```
[root@K8S-master01 2.2.8.1]# kubectl describe deploy nginx-deployment
Name:                   nginx-deployment
Namespace:              default
CreationTimestamp:      Thu, 24 Jan 2019 15:15:15 +0800
Labels:                 app=nginx
Annotations:            deployment.kubernetes.io/revision: 2
                        kubernetes.io/change-cause: kubectl set image deployment nginx-deployment nginx=nginx:1.9.1 --record=true
Selector:               app=nginx
Replicas:               3 desired | 3 updated | 3 total | 3 available | 0 unavailable
StrategyType:           RollingUpdate
MinReadySeconds:        0
RollingUpdateStrategy:  25% max unavailable, 25% max surge
Pod Template:
  Labels: app=nginx
  Containers:
   nginx:
    Image:        nginx:1.9.1
    Port:         80/TCP
    Host Port:    0/TCP
    Environment:  <none>
    Mounts:       <none>
  Volumes:        <none>
Conditions:
  Type           Status   Reason
  ----           ------   ------
  Available      True     MinimumReplicasAvailable
  Progressing    True     NewReplicaSetAvailable
OldReplicaSets:  <none>
NewReplicaSet:   nginx-deployment-6987cdb55b (3/3 replicas created)
Events:
  Type    Reason             Age    From                   Message
  ----    ------             ----   ----                   -------
  Normal  ScalingReplicaSet  36m    deployment-controller  Scaled up replica set nginx-deployment-5c689d88bb to 3
  Normal  ScalingReplicaSet  7m16s  deployment-controller  Scaled up replica set nginx-deployment-6987cdb55b to 1
  Normal  ScalingReplicaSet  5m18s  deployment-controller  Scaled down replica set nginx-deployment-5c689d88bb to 2
  Normal  ScalingReplicaSet  5m18s  deployment-controller  Scaled up replica set nginx-deployment-6987cdb55b to 2
  Normal  ScalingReplicaSet  4m35s  deployment-controller  Scaled down replica set nginx-deployment-5c689d88bb to 1
```

```
    Normal  ScalingReplicaSet  4m34s  deployment-controller  Scaled up replica
set nginx-deployment-6987cdb55b to 3
    Normal  ScalingReplicaSet  3m30s  deployment-controller  Scaled down
replica set nginx-deployment-5c689d88bb to 0
```

在 describe 中可以看出，第一次创建时，它创建了一个名为 nginx-deployment-5c689d88bb 的 ReplicaSet，并直接将其扩展为 3 个副本。更新部署时，它创建了一个新的 ReplicaSet，命名为 nginx-deployment-6987cdb55b，并将其副本数扩展为 1，然后将旧的 ReplicaSet 缩小为 2，这样至少可以有 2 个 Pod 可用，最多创建了 4 个 Pod。以此类推，使用相同的滚动更新策略向上和向下扩展新旧 ReplicaSet，最终新的 ReplicaSet 可以拥有 3 个副本，并将旧的 ReplicaSet 缩小为 0。

3. 回滚 Deployment

当新版本不稳定时，可以对其进行回滚操作，默认情况下，所有 Deployment 的 rollout 历史都保留在系统中，可以随时回滚。

假设我们又进行了几次更新：

```
    [root@K8S-master01 2.2.8.1]# kubectl set image deployment nginx-deployment
nginx=dotbalo/canary:v1 --record
    [root@K8S-master01 2.2.8.1]# kubectl set image deployment nginx-deployment
nginx=dotbalo/canary:v2 --record
```

使用 kubectl rollout history 查看部署历史：

```
    [root@K8S-master01 2.2.8.1]# kubectl rollout history
deployment/nginx-deployment
    deployment.extensions/nginx-deployment
    REVISION   CHANGE-CAUSE
    1          <none>
    2          kubectl set image deployment nginx-deployment nginx=nginx:1.9.1
--record=true
    3          kubectl set image deployment nginx-deployment nginx=dotbalo/canary:v1
--record=true
    4          kubectl set image deployment nginx-deployment nginx=dotbalo/canary:v2
--record=true
```

查看 Deployment 某次更新的详细信息，使用 --revision 指定版本号：

```
    [root@K8S-master01 2.2.8.1]# kubectl rollout history
deployment.v1.apps/nginx-deployment --revision=3
    deployment.apps/nginx-deployment with revision #3
    Pod Template:
      Labels:    app=nginx
        pod-template-hash=645959bf6b
      Annotations:  kubernetes.io/change-cause: kubectl set image deployment
nginx-deployment nginx=dotbalo/canary:v1 --record=true
      Containers:
       nginx:
        Image:     dotbalo/canary:v1
        Port:      80/TCP
        Host Port: 0/TCP
        Environment:<none>
        Mounts:    <none>
      Volumes:     <none>
```

使用 kubectl rollout undo 回滚到上一个版本：

```
[root@K8S-master01 2.2.8.1]# kubectl rollout undo
deployment.v1.apps/nginx-deployment
    deployment.apps/nginx-deployment
```

再次查看更新历史，发现 REVISION5 回到了 canary:v1：

```
[root@K8S-master01 2.2.8.1]# kubectl rollout history
deployment/nginx-deployment
    deployment.extensions/nginx-deployment
    REVISION   CHANGE-CAUSE
    1          <none>
    2          kubectl set image deployment nginx-deployment nginx=nginx:1.9.1
--record=true
    4          kubectl set image deployment nginx-deployment nginx=dotbalo/canary:v2
--record=true
    5          kubectl set image deployment nginx-deployment nginx=dotbalo/canary:v1
--record=true
```

使用 --to-revision 参数回到指定版本：

```
[root@K8S-master01 2.2.8.1]# kubectl rollout undo deployment/nginx-deployment
--to-revision=2
    deployment.extensions/nginx-deployment
```

4. 扩展 Deployment

当公司访问量变大，三个 Pod 已无法支撑业务时，可以对其进行扩展。

使用 kubectl scale 动态调整 Pod 的副本数，比如增加 Pod 为 5 个：

```
[root@K8S-master01 2.2.8.1]# kubectl scale
deployment.v1.apps/nginx-deployment --replicas=5
    deployment.apps/nginx-deployment scaled
```

查看 Pod，此时 Pod 已经变成了 5 个：

```
[root@K8S-master01 2.2.8.1]# kubectl get po
NAME                                READY   STATUS    RESTARTS   AGE
nginx-deployment-5f89547d9c-5r56b   1/1     Running   0          90s
nginx-deployment-5f89547d9c-htmn7   1/1     Running   0          25s
nginx-deployment-5f89547d9c-nwxs2   1/1     Running   0          99s
nginx-deployment-5f89547d9c-rpwlg   1/1     Running   0          25s
nginx-deployment-5f89547d9c-vlr5p   1/1     Running   0          95s
```

5. 暂停和恢复 Deployment 更新

Deployment 支持暂停更新，用于对 Deployment 进行多次修改操作。

使用 kubectl rollout pause 暂停 Deployment 更新：

```
[root@K8S-master01 2.2.8.1]# kubectl rollout pause
deployment/nginx-deployment
    deployment.extensions/nginx-deployment paused
```

然后对 Deployment 进行相关更新操作，比如更新镜像，然后对其资源进行限制：

```
[root@K8S-master01 2.2.8.1]# kubectl set image
deployment.v1.apps/nginx-deployment nginx=nginx:1.9.1
```

```
    deployment.apps/nginx-deployment image updated
    [root@K8S-master01 2.2.8.1]# kubectl set resources
deployment.v1.apps/nginx-deployment -c=nginx --limits=cpu=200m,memory=512Mi
    deployment.apps/nginx-deployment resource requirements updated
```

通过 rollout history 可以看到没有新的更新：

```
    [root@K8S-master01 2.2.8.1]# kubectl rollout history
deployment.v1.apps/nginx-deployment
    deployment.apps/nginx-deployment
    REVISION  CHANGE-CAUSE
    1         <none>
    5         kubectl set image deployment nginx-deployment nginx=dotbalo/canary:v1
--record=true
    7         kubectl set image deployment nginx-deployment nginx=dotbalo/canary:v2
--record=true
    8         kubectl set image deployment nginx-deployment nginx=dotbalo/canary:v2
--record=true
```

使用 kubectl rollout resume 恢复 Deployment 更新：

```
    [root@K8S-master01 2.2.8.1]# kubectl rollout resume
deployment.v1.apps/nginx-deployment
    deployment.apps/nginx-deployment resumed
```

可以查看到恢复更新的 Deployment 创建了一个新的 RS（复制集）：

```
    [root@K8S-master01 2.2.8.1]# kubectl get rs
    NAME                          DESIRED   CURRENT   READY   AGE
    nginx-deployment-57895845b8   5         5         4       11s
```

可以查看 Deployment 的 image（镜像）已经变为 nginx:1.9.1

```
    [root@K8S-master01 2.2.8.1]# kubectl describe deploy nginx-deployment
    Name:                   nginx-deployment
    Namespace:              default
    CreationTimestamp:      Thu, 24 Jan 2019 15:15:15 +0800
    Labels:                 app=nginx
    Annotations:            deployment.kubernetes.io/revision: 9
                            kubernetes.io/change-cause: kubectl set image deployment
nginx-deployment nginx=dotbalo/canary:v2 --record=true
    Selector:               app=nginx
    Replicas:               5 desired | 5 updated | 5 total | 5 available | 0
unavailable
    StrategyType:           RollingUpdate
    MinReadySeconds:        0
    RollingUpdateStrategy:  25% max unavailable, 25% max surge
    Pod Template:
      Labels: app=nginx
      Containers:
       nginx:
        Image:      nginx:1.9.1
        Port:       80/TCP
        Host Port:  0/TCP
```

6. 更新 Deployment 的注意事项

（1）清理策略

在默认情况下，revision 保留 10 个旧的 ReplicaSet，其余的将在后台进行垃圾回收，可以在 .spec.revisionHistoryLimit 设置保留 ReplicaSet 的个数。当设置为 0 时，不保留历史记录。

（2）更新策略

- .spec.strategy.type==Recreate，表示重建，先删掉旧的 Pod 再创建新的 Pod。
- .spec.strategy.type==RollingUpdate，表示滚动更新，可以指定 maxUnavailable 和 maxSurge 来控制滚动更新过程。
    - .spec.strategy.rollingUpdate.maxUnavailable，指定在回滚更新时最大不可用的 Pod 数量，可选字段，默认为 25%，可以设置为数字或百分比，如果 maxSurge 为 0，则该值不能为 0。
    - .spec.strategy.rollingUpdate.maxSurge 可以超过期望值的最大 Pod 数，可选字段，默认为 25%，可以设置成数字或百分比，如果 maxUnavailable 为 0，则该值不能为 0。

（3）Ready 策略

.spec.minReadySeconds 是可选参数，指定新创建的 Pod 应该在没有任何容器崩溃的情况下视为 Ready（就绪）状态的最小秒数，默认为 0，即一旦被创建就视为可用，通常和容器探针连用。

2.2.7 StatefulSet

StatefulSet（有状态集）常用于部署有状态的且需要有序启动的应用程序。

1. StatefulSet 的基本概念

StatefulSet 主要用于管理有状态应用程序的工作负载 API 对象。比如在生产环境中，可以部署 ElasticSearch 集群、MongoDB 集群或者需要持久化的 RabbitMQ 集群、Redis 集群、Kafka 集群和 ZooKeeper 集群等。

和 Deployment 类似，一个 StatefulSet 也同样管理着基于相同容器规范的 Pod。不同的是，StatefulSet 为每个 Pod 维护了一个粘性标识。这些 Pod 是根据相同的规范创建的，但是不可互换，每个 Pod 都有一个持久的标识符，在重新调度时也会保留，一般格式为 StatefulSetName-Number。比如定义一个名字是 Redis-Sentinel 的 StatefulSet，指定创建三个 Pod，那么创建出来的 Pod 名字就为 Redis-Sentinel-0、Redis-Sentinel-1、Redis-Sentinel-2。而 StatefulSet 创建的 Pod 一般使用 Headless Service（无头服务）进行通信，和普通的 Service 的区别在于 Headless Service 没有 ClusterIP，它使用的是 Endpoint 进行互相通信，Headless 一般的格式为：

statefulSetName-{0..N-1}.serviceName.namespace.svc.cluster.local。

说明：

- serviceName 为 Headless Service 的名字。
- 0..N-1 为 Pod 所在的序号，从 0 开始到 N-1。

- statefulSetName 为 StatefulSet 的名字。
- namespace 为服务所在的命名空间。
- .cluster.local 为 Cluster Domain（集群域）。

比如，一个 Redis 主从架构，Slave 连接 Master 主机配置就可以使用不会更改的 Master 的 Headless Service，例如 Redis 从节点（Slave）配置文件如下：

```
port 6379
slaveofredis-sentinel-master-ss-0.redis-sentinel-master-ss.public-service.svc.cluster.local 6379
tcp-backlog 511
timeout 0
tcp-keepalive 0
……
```

其中，redis-sentinel-master-ss-0.redis-sentinel-master-ss.public-service.svc.cluster.local 是 Redis Master 的 Headless Service。具体 Headless 可以参考 2.2.13 节。

2. 使用 StatefulSet

一般 StatefulSet 用于有以下一个或者多个需求的应用程序：

- 需要稳定的独一无二的网络标识符。
- 需要持久化数据。
- 需要有序的、优雅的部署和扩展。
- 需要有序的、自动滚动更新。

如果应用程序不需要任何稳定的标识符或者有序的部署、删除或者扩展，应该使用无状态的控制器部署应用程序，比如 Deployment 或者 ReplicaSet。

3. StatefulSet 的限制

StatefulSet 是 Kubernetes 1.9 版本之前的 beta 资源，在 1.5 版本之前的任何 Kubernetes 版本都没有。

Pod 所用的存储必须由 PersistentVolume Provisioner（持久化卷配置器）根据请求配置 StorageClass，或者由管理员预先配置。

为了确保数据安全，删除和缩放 StatefulSet 不会删除与 StatefulSet 关联的卷，可以手动选择性地删除 PVC 和 PV（关于 PV 和 PVC 请参考 2.2.12 节）。

StatefulSet 目前使用 Headless Service（无头服务）负责 Pod 的网络身份和通信，但需要创建此服务。

删除一个 StatefulSet 时，不保证对 Pod 的终止，要在 StatefulSet 中实现 Pod 的有序和正常终止，可以在删除之前将 StatefulSet 的副本缩减为 0。

4. StatefulSet 组件

定义一个简单的 StatefulSet 的示例如下：

```
apiVersion: v1
kind: Service
```

```yaml
metadata:
  name: nginx
  labels:
    app: nginx
spec:
  ports:
  - port: 80
    name: web
  clusterIP: None
  selector:
    app: nginx
---
apiVersion: apps/v1beta1
kind: StatefulSet
metadata:
  name: web
spec:
  serviceName: "nginx"
  replicas: 2
  template:
    metadata:
      labels:
        app: nginx
    spec:
      containers:
      - name: nginx
        image: nginx
        ports:
        - containerPort: 80
          name: web
        volumeMounts:
        - name: www
          mountPath: /usr/share/nginx/html
  volumeClaimTemplates:
  - metadata:
      name: www
    spec:
      accessModes: [ "ReadWriteOnce" ]
      storageClassName: "nginx-storage-class"
      resources:
        requests:
          storage: 1Gi
```

其中，

- kind: Service 定义了一个名字为 Nginx 的 Headless Service，创建的 Service 格式为 nginx-0.nginx.default.svc.cluster.local，其他的类似，因为没有指定 Namespace（命名空间），所以默认部署在 default。
- kind: StatefulSet 定义了一个名字为 web 的 StatefulSet，replicas 表示部署 Pod 的副本数，本实例为 2。
- volumeClaimTemplates 表示将提供稳定的存储 PV（持久化卷）作持久化，PV 可以是手动创建或者自动创建。在上述示例中，每个 Pod 将配置一个 PV，当 Pod 重新调度到某个节点上时，Pod 会重新挂载 volumeMounts 指定的目录（当前 StatefulSet 挂载到 /usr/share/nginx/html），当删除 Pod 或者 StatefulSet 时，不会删除 PV。

在 StatefulSet 中必须设置 Pod 选择器（.spec.selector）用来匹配其标签（.spec.template.metadata.labels）。在 1.8 版本之前，如果未配置该字段（.spec.selector），将被设置为默认值，在 1.8 版本之后，如果未指定匹配 Pod Selector，则会导致 StatefulSet 创建错误。

当 StatefulSet 控制器创建 Pod 时，它会添加一个标签 statefulset.kubernetes.io/pod-name，该标签的值为 Pod 的名称，用于匹配 Service。

5. 创建 StatefulSet

创建 StatefulSet 之前，需要提前创建 StatefulSet 持久化所用的 PersistentVolumes（持久化卷，以下简称 PV，也可以使用 emptyDir 不对数据进行保留），当然也可以使用动态方式自动创建 PV，关于 PV 将在 2.2.12 节进行详解，本节只作为演示使用，也可以先阅读 2.2.12 节进行了解。

本例使用 NFS 提供静态 PV，假如已有一台 NFS 服务器，IP 地址为 192.168.2.2，配置的共享目录如下：

```
[root@nfs web]# cat /etc/exports | tail -1
/nfs/web/ *(rw,sync,no_subtree_check,no_root_squash)
[root@nfs web]# exportfs -r
[root@nfs web]# systemctl reload nfs-server
[root@nfs web]# ls -l /nfs/web/
total 0
drwxr-xr-x 2 root root 6 Jan 31 17:22 nginx0
drwxr-xr-x 2 root root 6 Jan 31 17:22 nginx1
drwxr-xr-x 2 root root 6 Jan 31 17:22 nginx2
drwxr-xr-x 2 root root 6 Jan 31 17:22 nginx3
drwxr-xr-x 2 root root 6 Jan 31 17:22 nginx4
drwxr-xr-x 2 root root 6 Jan 31 17:22 nginx5
```

Nginx0-5 作为 StatefulSet Pod 的 PV 的数据存储目录，使用 PersistentVolume 创建 PV，文件如下：

```
apiVersion: v1
kind: PersistentVolume
metadata:
  name: pv-nginx-5
spec:
  capacity:
    storage: 1Gi
  accessModes:
    - ReadWriteOnce
  volumeMode: Filesystem
  persistentVolumeReclaimPolicy: Recycle
  storageClassName: "nginx-storage-class"
  nfs:
    # real share directory
    path: /nfs/web/nginx5
    # nfs real ip
    server: 192.168.2.2
```

具体参数的配置及其含义，可参考 2.2.12 节。

创建 PV：

```
[root@K8S-master01 2.2.7]# kubectl create -f web-pv.yaml
persistentvolume/pv-nginx-0 created
persistentvolume/pv-nginx-1 created
persistentvolume/pv-nginx-2 created
persistentvolume/pv-nginx-3 created
persistentvolume/pv-nginx-4 created
persistentvolume/pv-nginx-5 created
```

查看 PV：

```
[root@K8S-master01 2.2.7]# kubectl get pv
NAME          CAPACITY   ACCESS MODES   RECLAIM POLICY   STATUS      CLAIM    STORAGECLASS          REASON   AGE
   pv-nginx-0    1Gi        RWO            Recycle          Available            nginx-storage-class            26s
   pv-nginx-1    1Gi        RWO            Recycle          Available            nginx-storage-class            26s
   pv-nginx-2    1Gi        RWO            Recycle          Available            nginx-storage-class            26s
   pv-nginx-3    1Gi        RWO            Recycle          Available            nginx-storage-class            26s
   pv-nginx-4    1Gi        RWO            Recycle          Available            nginx-storage-class            26s
   pv-nginx-5    1Gi        RWO            Recycle          Available            nginx-storage-class            26s
```

创建 StatefulSet：

```
[root@K8S-master01 2.2.7]# kubectl create -f sts-web.yaml
service/nginx created
statefulset.apps/web created
[root@K8S-master01 2.2.7]# kubectl get sts
NAME   DESIRED   CURRENT   AGE
web    2         2         12s
[root@K8S-master01 2.2.7]# kubectl get svc
NAME         TYPE        CLUSTER-IP   EXTERNAL-IP   PORT(S)   AGE
kubernetes   ClusterIP   10.96.0.1    <none>        443/TCP   7d2h
nginx        ClusterIP   None         <none>        80/TCP    16s
[root@K8S-master01 2.2.7]# kubectl get po -l app=nginx
NAME    READY   STATUS    RESTARTS   AGE
web-0   1/1     Running   0          2m5s
web-1   1/1     Running   0          115s
```

查看 PVC 和 PV，可以看到 StatefulSet 创建的两个 Pod 的 PVC 已经和 PV 绑定成功：

```
[root@K8S-master01 2.2.7]# kubectl get pvc
NAME        STATUS   VOLUME       CAPACITY   ACCESS MODES   STORAGECLASS          AGE
www-web-0   Bound    pv-nginx-5   1Gi        RWO            nginx-storage-class   2m31s
www-web-1   Bound    pv-nginx-0   1Gi        RWO            nginx-storage-class   2m21s
[root@K8S-master01 2.2.7]# kubectl get pv
NAME   CAPACITY   ACCESS MODES   RECLAIM POLICY   STATUS   CLAIM
```

```
STORAGECLASS              REASON      AGE
    pv-nginx-0       1Gi         RWO              Recycle            Bound
default/www-web-1    nginx-storage-class                4m8s
    pv-nginx-1       1Gi         RWO              Recycle            Available
nginx-storage-class                    4m8s
    pv-nginx-2       1Gi         RWO              Recycle            Available
nginx-storage-class                    4m8s
    pv-nginx-3       1Gi         RWO              Recycle            Available
nginx-storage-class                    4m8s
    pv-nginx-4       1Gi         RWO              Recycle            Available
nginx-storage-class                    4m8s
    pv-nginx-5       1Gi         RWO              Recycle            Bound
default/www-web-0    nginx-storage-class                4m8s
```

6. 部署和扩展保障

Pod 的部署和扩展规则如下：

- 对于具有 N 个副本的 StatefulSet，将按顺序从 0 到 N-1 开始创建 Pod。
- 当删除 Pod 时，将按照 N-1 到 0 的反顺序终止。
- 在缩放 Pod 之前，必须保证当前的 Pod 是 Running（运行中）或者 Ready（就绪）。
- 在终止 Pod 之前，它所有的继任者必须是完全关闭状态。

StatefulSet 的 pod.Spec.TerminationGracePeriodSeconds 不应该指定为 0，设置为 0 对 StatefulSet 的 Pod 是极其不安全的做法，优雅地删除 StatefulSet 的 Pod 是非常有必要的，而且是安全的，因为它可以确保在 Kubelet 从 APIServer 删除之前，让 Pod 正常关闭。

当创建上面的 Nginx 实例时，Pod 将按 web-0、web-1、web-2 的顺序部署 3 个 Pod。在 web-0 处于 Running 或者 Ready 之前，web-1 不会被部署，相同的，web-2 在 web-1 未处于 Running 和 Ready 之前也不会被部署。如果在 web-1 处于 Running 和 Ready 状态时，web-0 变成 Failed（失败）状态，那么 web-2 将不会被启动，直到 web-0 恢复为 Running 和 Ready 状态。

如果用户将 StatefulSet 的 replicas 设置为 1，那么 web-2 将首先被终止，在完全关闭并删除 web-2 之前，不会删除 web-1。如果 web-2 终止并且完全关闭后，web-0 突然失败，那么在 web-0 未恢复成 Running 或者 Ready 时，web-1 不会被删除。

7. StatefulSet 扩容和缩容

和 Deployment 类似，可以通过更新 replicas 字段扩容/缩容 StatefulSet，也可以使用 kubectl scale 或者 kubectl patch 来扩容/缩容一个 StatefulSet。

（1）扩容

将上述创建的 sts 副本增加到 5 个（扩容之前必须保证有创建完成的静态 PV，动态 PV 和 emptyDir）：

```
[root@K8S-master01 2.2.7]# kubectl scale sts web --replicas=5
statefulset.apps/web scaled
```

查看 Pod 及 PVC 的状态：

```
[root@K8S-master01 2.2.7]# kubectl get pvc
    NAME          STATUS    VOLUME          CAPACITY    ACCESS MODES    STORAGECLASS
```

```
                                      AGE
    www-web-0   Bound    pv-nginx-0   1Gi         RWO          nginx-storage-class
2m54s
    www-web-1   Bound    pv-nginx-2   1Gi         RWO          nginx-storage-class
2m44s
    www-web-2   Bound    pv-nginx-5   1Gi         RWO          nginx-storage-class
112s
    www-web-3   Bound    pv-nginx-1   1Gi         RWO          nginx-storage-class
75s
    www-web-4   Bound    pv-nginx-3   1Gi         RWO          nginx-storage-class
49s
    [root@K8S-master01 2.2.7]# kubectl get po
    NAME    READY   STATUS    RESTARTS   AGE
    web-0   1/1     Running   0          2m58s
    web-1   1/1     Running   0          2m48s
    web-2   1/1     Running   0          116s
    web-3   1/1     Running   0          79s
    web-4   1/1     Running   0          53s
```

也可使用以下命令动态查看:

```
kubectl get pods -w -l app=nginx
```

(2) 缩容

在一个终端动态查看:

```
[root@K8S-master01 2.2.7]# kubectl get pods -w -l app=nginx
NAME    READY   STATUS    RESTARTS   AGE
web-0   1/1     Running   0          4m37s
web-1   1/1     Running   0          4m27s
web-2   1/1     Running   0          3m35s
web-3   1/1     Running   0          2m58s
web-4   1/1     Running   0          2m32s
```

在另一个终端将副本数改为 3:

```
[root@K8S-master01 ~]# kubectl patch sts web -p '{"spec":{"replicas":3}}'
statefulset.apps/web patched
```

此时可以看到第一个终端显示 web-4 和 web-3 的 Pod 正在被删除(或终止):

```
[root@K8S-master01 2.2.7]# kubectl get pods -w -l app=nginx
NAME    READY   STATUS        RESTARTS   AGE
web-0   1/1     Running       0          4m37s
web-1   1/1     Running       0          4m27s
web-2   1/1     Running       0          3m35s
web-3   1/1     Running       0          2m58s
web-4   1/1     Running       0          2m32s
web-0   1/1     Running       0          5m8s
web-0   1/1     Running       0          5m11s
web-4   1/1     Terminating   0          3m36s
web-4   0/1     Terminating   0          3m38s
web-4   0/1     Terminating   0          3m47s
web-4   0/1     Terminating   0          3m47s
web-3   1/1     Terminating   0          4m13s
web-3   0/1     Terminating   0          4m14s
```

```
    web-3    0/1    Terminating    0    4m22s
    web-3    0/1    Terminating    0    4m22s
```

查看状态，此时 PV 和 PVC 不会被删除：

```
[root@K8S-master01 2.2.7]# kubectl get po
NAME    READY   STATUS    RESTARTS   AGE
web-0   1/1     Running   0          7m11s
web-1   1/1     Running   0          7m1s
web-2   1/1     Running   0          6m9s
[root@K8S-master01 2.2.7]# kubectl get pvc
NAME        STATUS   VOLUME       CAPACITY   ACCESS MODES   STORAGECLASS          AGE
www-web-0   Bound    pv-nginx-0   1Gi        RWO            nginx-storage-class   7m15s
www-web-1   Bound    pv-nginx-2   1Gi        RWO            nginx-storage-class   7m5s
www-web-2   Bound    pv-nginx-5   1Gi        RWO            nginx-storage-class   6m13s
www-web-3   Bound    pv-nginx-1   1Gi        RWO            nginx-storage-class   5m36s
www-web-4   Bound    pv-nginx-3   1Gi        RWO            nginx-storage-class   5m10s
[root@K8S-master01 2.2.7]# kubectl get pv
NAME         CAPACITY   ACCESS MODES   RECLAIM POLICY   STATUS      CLAIM               STORAGECLASS          REASON   AGE
pv-nginx-0   1Gi        RWO            Recycle          Bound       default/www-web-0   nginx-storage-class            78m
pv-nginx-1   1Gi        RWO            Recycle          Bound       default/www-web-3   nginx-storage-class            78m
pv-nginx-2   1Gi        RWO            Recycle          Bound       default/www-web-1   nginx-storage-class            78m
pv-nginx-3   1Gi        RWO            Recycle          Bound       default/www-web-4   nginx-storage-class            78m
pv-nginx-4   1Gi        RWO            Recycle          Available                       nginx-storage-class            78m
pv-nginx-5   1Gi        RWO            Recycle          Bound       default/www-web-2   nginx-storage-class            78m
```

8. 更新策略

在 Kubernetes 1.7 以上的版本中，StatefulSet 的 .spec.updateStrategy 字段允许配置和禁用容器的自动滚动更新、标签、资源限制以及 StatefulSet 中 Pod 的注释等。

（1）On Delete 策略

OnDelete 更新策略实现了传统（1.7 版本之前）的行为，它也是默认的更新策略。当我们选择这个更新策略并修改 StatefulSet 的 .spec.template 字段时，StatefulSet 控制器不会自动更新 Pod，我们必须手动删除 Pod 才能使控制器创建新的 Pod。

（2）RollingUpdate 策略

RollingUpdate（滚动更新）更新策略会更新一个 StatefulSet 中所有的 Pod，采用与序号索引相反的顺序进行滚动更新。

比如 Patch 一个名称为 web 的 StatefulSet 来执行 RollingUpdate 更新：

```
[root@K8S-master01 2.2.7]# kubectl patch statefulset web -p
'{"spec":{"updateStrategy":{"type":"RollingUpdate"}}}'
statefulset.apps/web patched
```

查看更改后的 StatefulSet：

```
[root@K8S-master01 2.2.7]# kubectl get sts web -o yaml | grep -A 1
"updateStrategy"
    updateStrategy:
      type: RollingUpdate
```

然后改变容器的镜像进行滚动更新：

```
[root@K8S-master01 2.2.7]# kubectl patch statefulset web --type='json'
-p='[{"op": "replace", "path": "/spec/template/spec/containers/0/image",
"value":"dotbalo/canary:v1"}]'
statefulset.apps/web patched
```

如上所述，StatefulSet 里的 Pod 采用和序号相反的顺序更新。在更新下一个 Pod 前，StatefulSet 控制器会终止每一个 Pod 并等待它们变成 Running 和 Ready 状态。在当前顺序变成 Running 和 Ready 状态之前，StatefulSet 控制器不会更新下一个 Pod，但它仍然会重建任何在更新过程中发生故障的 Pod，使用它们当前的版本。已经接收到请求的 Pod 将会被恢复为更新的版本，没有收到请求的 Pod 则会被恢复为之前的版本。

在更新过程中可以使用 kubectl rollout status sts/<name> 来查看滚动更新的状态：

```
[root@K8S-master01 2.2.7]# kubectl rollout status sts/web
Waiting for 1 pods to be ready...
waiting for statefulset rolling update to complete 1 pods at revision
web-56b5798f76...
Waiting for 1 pods to be ready...
Waiting for 1 pods to be ready...
waiting for statefulset rolling update to complete 2 pods at revision
web-56b5798f76...
Waiting for 1 pods to be ready...
Waiting for 1 pods to be ready...
statefulset rolling update complete 3 pods at revision web-56b5798f76...
```

查看更新后的镜像：

```
[root@K8S-master01 2.2.7]# for p in 0 1 2; do kubectl get po web-$p --template
'{{range $i, $c := .spec.containers}}{{$c.image}}{{end}}'; echo; done
    dotbalo/canary:v1
    dotbalo/canary:v1
    dotbalo/canary:v1
```

（3）分段更新

StatefulSet 可以使用 RollingUpdate 更新策略的 partition 参数来分段更新一个 StatefulSet。分段更新将会使 StatefulSet 中其余的所有 Pod（序号小于分区）保持当前版本，只更新序号大于等于分区的 Pod，利用此特性可以简单实现金丝雀发布（灰度发布）或者分阶段推出新功能等。注：金丝雀发布是指在黑与白之间能够平滑过渡的一种发布方式。

比如我们定义一个分区"partition":3，可以使用 patch 直接对 StatefulSet 进行设置：

```
# kubectl patch statefulset web -p
```

```
'{"spec":{"updateStrategy":{"type":"RollingUpdate","rollingUpdate":{"partition":3}}}}'
    statefulset "web" patched
```

然后再次 patch 改变容器的镜像：

```
# kubectl patch statefulset web --type='json' -p='[{"op": "replace", "path": "/spec/template/spec/containers/0/image", "value":"K8S.gcr.io/nginx-slim:0.7"}]'
    statefulset "web" patched
```

删除 Pod 触发更新：

```
kubectl delete po web-2
pod "web-2" deleted
```

此时，因为 Podweb-2 的序号小于分区 3，所以 Pod 不会被更新，还是会使用以前的容器恢复 Pod。

将分区改为 2，此时会自动更新 web-2（因为之前更改了更新策略），但是不会更新 web-0 和 web-1：

```
# kubectl patch statefulset web -p '{"spec":{"updateStrategy":{"type":"RollingUpdate","rollingUpdate":{"partition":2}}}}'
    statefulset "web" patched
```

按照上述方式，可以实现分阶段更新，类似于灰度/金丝雀发布。查看最终的结果如下：

```
[root@K8S-master01 2.2.7]# for p in 0 1 2; do kubectl get po web-$p --template '{{range $i, $c := .spec.containers}}{{$c.image}}{{end}}'; echo; done
dotbalo/canary:v1
dotbalo/canary:v1
dotbalo/canary:v2
```

9. 删除 StatefulSet

删除 StatefulSet 有两种方式，即级联删除和非级联删除。使用非级联方式删除 StatefulSet 时，StatefulSet 的 Pod 不会被删除；使用级联删除时，StatefulSet 和它的 Pod 都会被删除。

（1）非级联删除

使用 kubectldeletestsxxx 删除 StatefulSet 时，只需提供 --cascade=false 参数，就会采用非级联删除，此时删除 StatefulSet 不会删除它的 Pod：

```
[root@K8S-master01 2.2.7]# kubectl get po
NAME    READY   STATUS    RESTARTS   AGE
web-0   1/1     Running   0          16m
web-1   1/1     Running   0          16m
web-2   1/1     Running   0          11m
You have new mail in /var/spool/mail/root
[root@K8S-master01 2.2.7]# kubectl delete statefulset web --cascade=false
statefulset.apps "web" deleted
[root@K8S-master01 2.2.7]# kubectl get sts
No resources found.
[root@K8S-master01 2.2.7]# kubectl get po
NAME    READY   STATUS    RESTARTS   AGE
```

```
web-0    1/1    Running    0    16m
web-1    1/1    Running    0    16m
web-2    1/1    Running    0    11m
```

由于此时删除了 StatefulSet，因此单独删除 Pod 时，不会被重建：

```
[root@K8S-master01 2.2.7]# kubectl get po
NAME     READY  STATUS     RESTARTS  AGE
web-0    1/1    Running    0         16m
web-1    1/1    Running    0         16m
web-2    1/1    Running    0         11m
[root@K8S-master01 2.2.7]# kubectl delete po web-0
pod "web-0" deleted
[root@K8S-master01 2.2.7]# kubectl get po
NAME     READY  STATUS     RESTARTS  AGE
web-1    1/1    Running    0         18m
web-2    1/1    Running    0         12m
```

当再次创建此 StatefulSet 时，web-0 会被重新创建，web-1 由于已经存在而不会被再次创建，因为最初此 StatefulSet 的 replicas 是 2，所以 web-2 会被删除，如下（忽略 AlreadyExists 错误）：

```
[root@K8S-master01 2.2.7]# kubectl create -f sts-web.yaml
statefulset.apps/web created
Error from server (AlreadyExists): error when creating "sts-web.yaml": services "nginx" already exists
[root@K8S-master01 2.2.7]# kubectl get po
NAME     READY  STATUS     RESTARTS  AGE
web-0    1/1    Running    0         32s
web-1    1/1    Running    0         19m
```

（2）级联删除

省略 --cascade=false 参数即为级联删除：

```
[root@K8S-master01 2.2.7]# kubectl delete statefulset web
statefulset.apps "web" deleted
[root@K8S-master01 2.2.7]# kubectl get po
No resources found.
```

也可以使用 -f 参数直接删除 StatefulSet 和 Service（此文件将 sts 和 svc 写在了一起）：

```
[root@K8S-master01 2.2.7]# kubectl delete -f sts-web.yaml
service "nginx" deleted
Error from server (NotFound): error when deleting "sts-web.yaml": statefulsets.apps "web" not found
[root@K8S-master01 2.2.7]#
```

2.2.8 DaemonSet

DaemonSet（守护进程集）和守护进程类似，它在符合匹配条件的节点上均部署一个 Pod。

1. 什么是 DaemonSet

DaemonSet 确保全部（或者某些）节点上运行一个 Pod 副本。当有新节点加入集群时，也会为它们新增一个 Pod。当节点从集群中移除时，这些 Pod 也会被回收，删除 DaemonSet 将会删除它创建的所有 Pod。

使用 DaemonSet 的一些典型用法：

- 运行集群存储 daemon（守护进程），例如在每个节点上运行 Glusterd、Ceph 等。
- 在每个节点运行日志收集 daemon，例如 Fluentd、Logstash。
- 在每个节点运行监控 daemon，比如 Prometheus Node Exporter、Collectd、Datadog 代理、New Relic 代理或 Ganglia gmond。

2. 编写 DaemonSet 规范

创建一个 DaemonSet 的内容大致如下，比如创建一个 fluentd 的 DaemonSet：

```yaml
apiVersion: apps/v1
kind: DaemonSet
metadata:
  name: fluentd-es-v2.0.4
  namespace: logging
  labels:
    K8S-app: fluentd-es
    version: v2.0.4
    kubernetes.io/cluster-service: "true"
    addonmanager.kubernetes.io/mode: Reconcile
spec:
  selector:
    matchLabels:
      K8S-app: fluentd-es
      version: v2.0.4
  template:
    metadata:
      labels:
        K8S-app: fluentd-es
        kubernetes.io/cluster-service: "true"
        version: v2.0.4
      # This annotation ensures that fluentd does not get evicted if the node
      # supports critical pod annotation based priority scheme.
      # Note that this does not guarantee admission on the nodes (#40573).
      annotations:
        scheduler.alpha.kubernetes.io/critical-pod: ''
        seccomp.security.alpha.kubernetes.io/pod: 'docker/default'
    spec:
      serviceAccountName: fluentd-es
      containers:
      - name: fluentd-es
        image: K8S.gcr.io/fluentd-elasticsearch:v2.0.4
        env:
        - name: FLUENTD_ARGS
          value: --no-supervisor -q
        resources:
          limits:
            memory: 500Mi
          requests:
            cpu: 100m
            memory: 200Mi
        volumeMounts:
```

```yaml
        - name: varlog
          mountPath: /var/log
        - name: varlibdockercontainers
          mountPath: /var/lib/docker/containers
          readOnly: true
        - name: config-volume
          mountPath: /etc/fluent/config.d
      nodeSelector:
        beta.kubernetes.io/fluentd-ds-ready: "true"
      terminationGracePeriodSeconds: 30
      volumes:
      - name: varlog
        hostPath:
          path: /var/log
      - name: varlibdockercontainers
        hostPath:
          path: /var/lib/docker/containers
      - name: config-volume
        configMap:
          name: fluentd-es-config-v0.1.4
```

（1）必需字段

和其他所有 Kubernetes 配置一样，DaemonSet 需要 apiVersion、kind 和 metadata 字段，同时也需要一个 .spec 配置段。

（2）Pod 模板

.spec 唯一需要的字段是 .spec.template。.spec.template 是一个 Pod 模板，它与 Pod 具有相同的配置方式，但它不具有 apiVersion 和 kind 字段。

除了 Pod 必需的字段外，在 DaemonSet 中的 Pod 模板必须指定合理的标签。

在 DaemonSet 中的 Pod 模板必须具有一个 RestartPolicy，默认为 Always。

（3）Pod Selector

.spec.selector 字段表示 Pod Selector，它与其他资源的 .spec.selector 的作用相同。

.spec.selector 表示一个对象，它由如下两个字段组成：

- matchLabels，与 ReplicationController 的 .spec.selector 的作用相同，用于匹配符合条件的 Pod。
- matchExpressions，允许构建更加复杂的 Selector，可以通过指定 key、value 列表以及与 key 和 value 列表相关的操作符。

如果上述两个字段都指定时，结果表示的是 AND 关系（逻辑与的关系）。

.spec.selector 必须与 .spec.template.metadata.labels 相匹配。如果没有指定，默认是等价的，如果它们的配置不匹配，则会被 API 拒绝。

（4）指定节点部署 Pod

如果指定了 .spec.template.spec.nodeSelector，DaemonSet Controller 将在与 Node Selector（节点选择器）匹配的节点上创建 Pod，比如部署在磁盘类型为 ssd 的节点上（需要提前给节点定义标签 Label）：

```yaml
containers:
- name: nginx
  image: nginx
  imagePullPolicy: IfNotPresent
nodeSelector:
  disktype: ssd
```

> **提示**
>
> Node Selector 同样适用于其他 Controller。

3. 创建 DaemonSet

在生产环境中，公司业务的应用程序一般无须使用 DaemonSet 部署，一般情况下只有像 Fluentd（日志收集）、Ingress（集群服务入口）、Calico（集群网络组件）、Node-Exporter（监控数据采集）等才需要使用 DaemonSet 部署到每个节点。本节只演示 DaemonSet 的使用。

比如创建一个 nginxingress：

```
[root@K8S-master01 2.2.8]# pwd
/root/chap02/2.2.8
[root@K8S-master01 2.2.8]# kubectl create -f nginx-ds.yaml
namespace/ingress-nginx created
configmap/nginx-configuration created
configmap/tcp-services created
configmap/udp-services created
serviceaccount/nginx-ingress-serviceaccount created
clusterrole.rbac.authorization.K8S.io/nginx-ingress-clusterrole created
role.rbac.authorization.K8S.io/nginx-ingress-role created
rolebinding.rbac.authorization.K8S.io/nginx-ingress-role-nisa-binding created
clusterrolebinding.rbac.authorization.K8S.io/nginx-ingress-clusterrole-nisa-binding created
daemonset.extensions/nginx-ingress-controller created
```

此时会在每个节点创建一个 Pod：

```
[root@K8S-master01 2.2.8]# kubectl get po -n ingress-nginx
NAME                             READY   STATUS    RESTARTS   AGE
nginx-ingress-controller-fjkf2   1/1     Running   0          44s
nginx-ingress-controller-gfmcv   1/1     Running   0          44s
nginx-ingress-controller-j89qc   1/1     Running   0          44s
nginx-ingress-controller-sqsk2   1/1     Running   0          44s
nginx-ingress-controller-tgdt6   1/1     Running   0          44s
[root@K8S-master01 2.2.8]# kubectl get po -n ingress-nginx -o wide
NAME                             READY   STATUS    RESTARTS   AGE
   IP              NODE            NOMINATED NODE
nginx-ingress-controller-fjkf2   1/1     Running   0          50s
192.168.20.30    K8S-node01      <none>
nginx-ingress-controller-gfmcv   1/1     Running   0          50s
192.168.20.21    K8S-master02    <none>
nginx-ingress-controller-j89qc   1/1     Running   0          50s
192.168.20.22    K8S-master03    <none>
nginx-ingress-controller-sqsk2   1/1     Running   0          50s
192.168.20.31    K8S-node02      <none>
nginx-ingress-controller-tgdt6   1/1     Running   0          50s
```

```
192.168.20.20    K8S-master01    <none>
```

> **注 意**
>
> 因为笔者的 Master 节点删除了 Taint（Taint 和 Toleration 见 2.2.18），所以也能部署 Ingress 或者其他 Pod，在生产环境下，在 Master 节点最好除了系统组件外不要部署其他 Pod。

4．更新和回滚 DaemonSet

如果修改了节点标签（Label），DaemonSet 将立刻向新匹配上的节点添加 Pod，同时删除不能匹配的节点上的 Pod。

在 Kubernetes 1.6 以后的版本中，可以在 DaemonSet 上执行滚动更新，未来的 Kubernetes 版本将支持节点的可控更新。

DaemonSet 滚动更新可参考：https://kubernetes.io/docs/tasks/manage-daemon/update-daemon-set/。

DaemonSet 更新策略和 StatefulSet 类似，也有 OnDelete 和 RollingUpdate 两种方式。

查看上一节创建的 DaemonSet 更新方式：

```
[root@K8S-master01 2.2.8]# kubectl get ds/nginx-ds -o
go-template='{{.spec.updateStrategy.type}}{{"\n"}}'
    RollingUpdate
```

> **提 示**
>
> 如果是其他 DaemonSet，请确保更新策略是 RollingUpdate（滚动更新）。

（1）命令式更新

```
kubectl edit ds/<daemonset-name>
kubectl patch ds/<daemonset-name> -p=<strategic-merge-patch>
```

（2）更新镜像

```
kubectl set image
ds/<daemonset-name><container-name>=<container-new-image>--record=true
```

（3）查看更新状态

```
kubectl rollout status ds/<daemonset-name>
```

（4）列出所有修订版本

```
kubectl rollout history daemonset <daemonset-name>
```

（5）回滚到指定 revision

```
kubectl rollout undo daemonset <daemonset-name> --to-revision=<revision>
```

DaemonSet 的更新和回滚与 Deployment 类似，此处不再演示。

2.2.9 ConfigMap

一般用 ConfigMap 管理一些程序的配置文件或者 Pod 变量，比如 Nginx 配置、MavenSetting

配置文件等。

1. 什么是 ConfigMap

ConfigMap 是一个将配置文件、命令行参数、环境变量、端口号和其他配置绑定到 Pod 的容器和系统组件。ConfigMaps 允许将配置与 Pod 和组件分开，这有助于保持工作负载的可移植性，使配置更易于更改和管理。比如在生产环境中，可以将 Nginx、Redis 等应用的配置文件存储在 ConfigMap 上，然后将其挂载即可使用。

相对于 Secret，ConfigMap 更倾向于存储和共享非敏感、未加密的配置信息，如果要在集群中使用敏感信息，最好使用 Secret。

2. 创建 ConfigMap

可以使用 kubectl create configmap 命令从目录、文件或字符值创建 ConfigMap：

```
kubectl create configmap <map-name><data-source>
```

说明：

- map-name，ConfigMap 的名称。
- data-source，数据源，数据的目录、文件或字符值。

数据源对应于 ConfigMap 中的键-值对（key-value pair），其中，

- key：文件名或密钥。
- value：文件内容或字符值。

（1）从目录创建 ConfigMap

可以使用 kubectl create configmap 命令从同一个目录中的多个文件创建 ConfigMap。

创建一个配置文件目录并且下载两个文件作为测试配置文件：

```
mkdir -p configure-pod-container/configmap/kubectl/

wget https://K8S.io/docs/tasks/configure-pod-container/configmap/kubectl/game.properties -O configure-pod-container/configmap/kubectl/game.properties

wget https://K8S.io/docs/tasks/configure-pod-container/configmap/kubectl/ui.properties -O configure-pod-container/configmap/kubectl/ui.properties
```

创建 ConfigMap，默认创建在 default 命名空间下，可以使用-n 更改 NameSpace（命名空间）：

```
[root@K8S-master01 ~]# kubectl create configmap game-config
--from-file=configure-pod-container/configmap/kubectl/
configmap/game-config created
```

查看当前的 ConfigMap：

```
[root@K8S-master01 ~]# kubectl describe configmaps game-config
Name:         game-config
Namespace:    default
Labels:       <none>
```

```
Annotations:  <none>

Data
====
game.properties:
----
enemies=aliens
lives=3
enemies.cheat=true
enemies.cheat.level=noGoodRotten
secret.code.passphrase=UUDDLRLRBABAS
secret.code.allowed=true
secret.code.lives=30
ui.properties:
----
color.good=purple
color.bad=yellow
allow.textmode=true
how.nice.to.look=fairlyNice

Events:  <none>
```

可以看到，ConfigMap 的内容与测试的配置文件内容一致：

```
[root@K8S-master01 ~]# cat configure-pod-container/configmap/kubectl/game.properties
   enemies=aliens
   lives=3
   enemies.cheat=true
   enemies.cheat.level=noGoodRotten
   secret.code.passphrase=UUDDLRLRBABAS
   secret.code.allowed=true
   secret.code.lives=30[root@K8S-master01 ~]# cat configure-pod-container/configmap/kubectl/ui.properties
   color.good=purple
   color.bad=yellow
   allow.textmode=true
   how.nice.to.look=fairlyNice
```

（2）从文件创建 ConfigMap

可以使用 kubectl create configmap 命令从单个文件或多个文件创建 ConfigMap。

例如以 configure-pod-container/configmap/kubectl/game.properties 文件建立 ConfigMap：

```
[root@K8S-master01 ~]# kubectl create configmap game-config-2
--from-file=configure-pod-container/configmap/kubectl/game.properties
configmap/game-config-2 created
```

查看当前的 ConfigMap：

```
[root@K8S-master01 ~]# kubectl get cm game-config-2
NAME             DATA   AGE
game-config-2    1      38s
```

也可以使用 --from-file 多次传入参数以从多个数据源创建 ConfigMap：

```
[root@K8S-master01 ~]# kubectl create configmap game-config-3
```

```
--from-file=configure-pod-container/configmap/kubectl/game.properties
--from-file=configure-pod-container/configmap/kubectl/ui.properties
   configmap/game-config-3 created
```

查看当前的 ConfigMap：

```
[root@K8S-master01 ~]# kubectl get cm game-config-3 -oyaml
apiVersion: v1
data:
  game.properties: |-
    enemies=aliens
    lives=3
    enemies.cheat=true
    enemies.cheat.level=noGoodRotten
    secret.code.passphrase=UUDDLRLRBABAS
    secret.code.allowed=true
    secret.code.lives=30
  ui.properties: |
    color.good=purple
    color.bad=yellow
    allow.textmode=true
    how.nice.to.look=fairlyNice
kind: ConfigMap
metadata:
  creationTimestamp: 2019-02-11T08:33:34Z
  name: game-config-3
  namespace: default
  resourceVersion: "4266928"
  selfLink: /api/v1/namespaces/default/configmaps/game-config-3
  uid: b88eea8b-2dd7-11e9-9180-000c293ad492
```

（3）从 ENV 文件创建 ConfigMap

可以使用 --from-env-file 从 ENV 文件创建 ConfigMap。

首先创建/下载一个测试文件，文件内容为 key=value 的格式：

```
[root@K8S-master01 ~]# wget https://K8S.io/docs/tasks/configure-pod-container/configmap/kubectl/game-env-file.properties -O configure-pod-container/configmap/kubectl/game-env-file.properties

[root@K8S-master01 ~]# cat configure-pod-container/configmap/kubectl/game-env-file.properties
enemies=aliens
lives=3
allowed="true"

# This comment and the empty line above it are ignored
```

创建 ConfigMap：

```
[root@K8S-master01 ~]# kubectl create configmap game-config-env-file \
--from-env-file=configure-pod-container/configmap/kubectl/game-env-file.properties
   configmap/game-config-env-file created
```

查看当前的 ConfigMap：

```
[root@K8S-master01 ~]# kubectl get configmap game-config-env-file -o yaml
apiVersion: v1
data:
  allowed: '"true"'
  enemies: aliens
  lives: "3"
kind: ConfigMap
metadata:
  creationTimestamp: 2019-02-11T08:40:17Z
  name: game-config-env-file
  namespace: default
  resourceVersion: "4267912"
  selfLink: /api/v1/namespaces/default/configmaps/game-config-env-file
  uid: a84ccd32-2dd8-11e9-90e9-000c293bfe27
```

注　意

如果使用 --from-env-file 多次传递参数以从多个数据源创建 ConfigMap 时，仅最后一个 ENV 生效。

（4）自定义 data 文件名创建 ConfigMap

可以使用以下命令自定义文件名：

```
kubectl create configmap game-config-3
--from-file=<my-key-name>=<path-to-file>
```

比如将 game.properties 文件定义为 game-special-key：

```
[root@K8S-master01 ~]# kubectl create configmap game-config-4
--from-file=game-special-key=configure-pod-container/configmap/kubectl/game.properties
configmap/game-config-4 created
[root@K8S-master01 ~]# kubectl get configmaps game-config-4 -o yaml
apiVersion: v1
data:
  game-special-key: |-
    enemies=aliens
    lives=3
    enemies.cheat=true
    enemies.cheat.level=noGoodRotten
    secret.code.passphrase=UUDDLRLRBABAS
    secret.code.allowed=true
    secret.code.lives=30
kind: ConfigMap
metadata:
  creationTimestamp: 2019-02-11T08:46:08Z
  name: game-config-4
  namespace: default
  resourceVersion: "4268642"
  selfLink: /api/v1/namespaces/default/configmaps/game-config-4
  uid: 797d269a-2dd9-11e9-90e9-000c293bfe27
```

（5）从字符值创建 ConfigMaps

可以使用 kubectl create configmap 与 --from-literal 参数来定义命令行的字符值：

```
[root@K8S-master01 ~]# kubectl create configmap special-config
--from-literal=special.how=very --from-literal=special.type=charm
configmap/special-config created
[root@K8S-master01 ~]# kubectl get cm special-config -o yaml
apiVersion: v1
data:
  special.how: very
  special.type: charm
kind: ConfigMap
metadata:
  creationTimestamp: 2019-02-11T08:49:28Z
  name: special-config
  namespace: default
  resourceVersion: "4269314"
  selfLink: /api/v1/namespaces/default/configmaps/special-config
  uid: f0dbb926-2dd9-11e9-8f6f-000c298bf023
```

3. ConfigMap 实践

本节主要讲解 ConfigMap 的一些常见使用方法，比如通过单个 ConfigMap 定义环境变量、通过多个 ConfigMap 定义环境变量和将 ConfigMap 作为卷使用等。

（1）使用单个 ConfigMap 定义容器环境变量

首先在 ConfigMap 中将环境变量定义为键-值对（key-value pair）：

```
kubectl create configmap special-config --from-literal=special.how=very
```

然后，将 ConfigMap 中定义的值 special.how 分配给 Pod 的环境变量 SPECIAL_LEVEL_KEY：

```
apiVersion: v1
kind: Pod
metadata:
  name: dapi-test-pod
spec:
  containers:
    - name: test-container
      image: busybox
      command: [ "/bin/sh", "-c", "env" ]
      env:
        # Define the environment variable
    - name: SPECIAL_LEVEL_KEY
          valueFrom:
            configMapKeyRef:
              # The ConfigMap containing the value you want to assign to
SPECIAL_LEVEL_KEY
    name: special-config
              # Specify the key associated with the value
    key: special.how
  restartPolicy: Never
```

（2）使用多个 ConfigMap 定义容器环境变量

首先定义两个或多个 ConfigMap：

```yaml
apiVersion: v1
kind: ConfigMap
metadata:
  name: special-config
  namespace: default
data:
  special.how: very
apiVersion: v1
kind: ConfigMap
metadata:
  name: env-config
  namespace: default
data:
  log_level: INFO
```

然后，在 Pod 中引用 ConfigMap：

```yaml
apiVersion: v1
kind: Pod
metadata:
  name: dapi-test-pod
spec:
  containers:
    - name: test-container
      image: busybox
      command: [ "/bin/sh", "-c", "env" ]
      env:
- name: SPECIAL_LEVEL_KEY
        valueFrom:
          configMapKeyRef:
            name: special-config
            key: special.how
        - name: LOG_LEVEL
          valueFrom:
            configMapKeyRef:
              name: env-config
              key: log_level
  restartPolicy: Never
```

（3）将 ConfigMap 中所有的键-值对配置为容器的环境变量

创建含有多个键-值对的 ConfigMap：

```yaml
apiVersion: v1
kind: ConfigMap
metadata:
  name: special-config
  namespace: default
data:
  SPECIAL_LEVEL: very
  SPECIAL_TYPE: charm
```

使用 envFrom 将 ConfigMap 所有的键-值对作为容器的环境变量，其中 ConfigMap 中的键作为 Pod 中的环境变量的名称：

```yaml
apiVersion: v1
kind: Pod
```

```yaml
metadata:
  name: dapi-test-pod
spec:
  containers:
    - name: test-container
      image: busybox
      command: [ "/bin/sh", "-c", "env" ]
  envFrom:
    - configMapRef:
        name: special-config
  restartPolicy: Never
```

（4）将 ConfigMap 添加到卷

大部分情况下，ConfigMap 定义的都是配置文件，不是环境变量，因此需要将 ConfigMap 中的文件（一般为--from-file 创建）挂载到 Pod 中，然后 Pod 中的容器就可引用，此时可以通过 volume 进行挂载。

例如，将名称为 special-config 的 ConfigMap，挂载到容器的/etc/config/目录下：

```yaml
apiVersion: v1
kind: Pod
metadata:
  name: dapi-test-pod
spec:
  containers:
    - name: test-container
      image: busybox
      command: [ "/bin/sh", "-c", "ls /etc/config/" ]
      volumeMounts:
      - name: config-volume
        mountPath: /etc/config
  volumes:
    - name: config-volume
      configMap:
        # Provide the name of the ConfigMap containing the files you want
        # to add to the container
        name: special-config
  restartPolicy: Never
```

此时 Pod 运行，会执行 command 的命令，即执行 ls /etc/config/

```
special.level
special.type
```

> **注　意**
>
> /etc/config/会被覆盖。

（5）将 ConfigMap 添加到卷并指定文件名

使用 path 字段可以指定 ConfigMap 挂载的文件名，比如将 special.level 挂载到/etc/config，并指定名称为 keys：

```yaml
apiVersion: v1
kind: Pod
```

```yaml
metadata:
  name: dapi-test-pod
spec:
  containers:
    - name: test-container
      image: busybox
      command: [ "/bin/sh","-c","cat /etc/config/keys" ]
      volumeMounts:
        - name: config-volume
          mountPath: /etc/config
  volumes:
    - name: config-volume
      configMap:
        name: special-config
        items:
          - key: special.level
            path: keys
  restartPolicy: Never
```

此时启动 Pod 时会打印：very

（6）指定特定路径和文件权限

方式和 Secret 类似，可参考 2.2.10.4 节的内容。

4. ConfigMap 限制

（1）必须先创建 ConfigMap 才能在 Pod 中引用它，如果 Pod 引用的 ConfigMap 不存在，Pod 将无法启动。

（2）Pod 引用的键必须存在于 ConfigMap 中，否则 Pod 无法启动。

（3）使用 envFrom 配置容器环境变量时，默认会跳过被视为无效的键，但是不影响 Pod 启动，无效的变量会记录在事件日志中，如下：

```
kubectl get events
     LASTSEEN FIRSTSEEN COUNT NAME           KIND  SUBOBJECT  TYPE
REASON                       SOURCE              MESSAGE
     0s       0s        1     dapi-test-pod  Pod              Warning
InvalidEnvironmentVariableNames  {kubelet, 127.0.0.1}  Keys [1badkey,
2alsobad] from the EnvFrom configMap default/myconfig were skipped since they
are considered invalid environment variable names.
```

（4）ConfigMap 和引用它的 Pod 需要在同一个命名空间。

2.2.10 Secret

Secret 对象类型用来保存敏感信息，例如密码、令牌和 SSH Key，将这些信息放在 Secret 中比较安全和灵活。用户可以创建 Secret 并且引用到 Pod 中，比如使用 Secret 初始化 Redis、MySQL 等密码。

1. 创建 Secret

创建 Secret 的方式有很多，比如使用命令行 Kubelet 或者使用 Yaml/Json 文件创建等。

（1）使用 Kubectl 创建 Secret

假设有些 Pod 需要访问数据库，可以将账户密码存储在 username.txt 和 password.txt 文件里，然后以文件的形式创建 Secret 供 Pod 使用。

创建账户信息文件：

```
[root@K8S-master01 ~]# echo -n "admin" > ./username.txt
[root@K8S-master01 ~]# echo -n "1f2d1e2e67df" > ./password.txt
```

以文件 username.txt 和 password.txt 创建 Secret：

```
[root@K8S-master01 ~]# kubectl create secret generic db-user-pass
--from-file=./username.txt --from-file=./password.txt
secret/db-user-pass created
```

查看 Secret：

```
[root@K8S-master01 ~]# kubectl get secrets db-user-pass
NAME            TYPE     DATA   AGE
db-user-pass    Opaque   2      33s
[root@K8S-master01 ~]# kubectl describe secrets/db-user-pass
Name:         db-user-pass
Namespace:    default
Labels:       <none>
Annotations:  <none>

Type: Opaque

Data
====
password.txt:  12 bytes
username.txt:  0 bytes
```

默认情况下，get 和 describe 命令都不会显示文件的内容，这是为了防止 Secret 中的内容被意外暴露。可以参考 2.2.10.2 一节的方式解码 Secret。

（2）手动创建 Secret

手动创建 Secret，因为每一项内容必须是 base64 编码，所以要先对其进行编码：

```
[root@K8S-master01 ~]# echo -n "admin" | base64
YWRtaW4=
[root@K8S-master01 ~]# echo -n "1f2d1e2e67df" | base64
MWYyZDFlMmU2N2Rm
```

然后，创建一个文件，内容如下：

```
[root@K8S-master01 ~]# cat db-user-secret.yaml
apiVersion: v1
kind: Secret
metadata:
  name: mysecret
type: Opaque
data:
  username: YWRtaW4=
  password: MWYyZDFlMmU2N2Rm
```

最后，使用该文件创建一个 Secret：

```
[root@K8S-master01 ~]# kubectl create -f db-user-secret.yaml
secret/mysecret created
```

2. 解码 Secret

Secret 被创建后，会以加密的方式存储于 Kubernetes 集群中，可以对其进行解码获取内容。首先以 yaml 的形式获取刚才创建的 Secret：

```
[root@K8S-master01 ~]# kubectl get secret mysecret -o yaml
apiVersion: v1
data:
  password: MWYyZDFlMmU2N2Rm
  username: YWRtaW4=
kind: Secret
metadata:
  creationTimestamp: 2019-02-09T03:16:19Z
  name: mysecret
  namespace: default
  resourceVersion: "3811354"
  selfLink: /api/v1/namespaces/default/secrets/mysecret
  uid: 11e49e9f-2c19-11e9-8f6f-000c298bf023
type: Opaque
```

然后通过 --decode 解码 Secret：

```
[root@K8S-master01 ~]# echo "MWYyZDFlMmU2N2Rm" | base64 --decode
1f2d1e2e67df
```

3. 使用 Secret

Secret 可以作为数据卷被挂载，或作为环境变量以供 Pod 的容器使用。

（1）在 Pod 中使用 Secret

在 Pod 中的 volume 里使用 Secret：

①首先创建一个 Secret 或者使用已有的 Secret，多个 Pod 可以引用同一个 Secret。

②在 spec.volumes 下增加一个 volume，命名随意，spec.volumes.secret.secretName 必须和 Secret 对象的名字相同，并且在同一个 Namespace 中。

③将 spec.containers.volumeMounts 加到需要用到该 Secret 的容器中，并且设置 spec.containers.volumeMounts.readOnly = true。

④使用 spec.containers.volumeMounts.mountPath 指定 Secret 挂载目录。

例如，将名字为 mysecret 的 Secret 挂载到 Pod 中的 /etc/foo：

```
apiVersion: v1
kind: Pod
metadata:
  name: mypod
spec:
  containers:
  - name: mypod
    image: redis
```

```yaml
    volumeMounts:
    - name: foo
      mountPath: "/etc/foo"
      readOnly: true
  volumes:
  - name: foo
    secret:
      secretName: mysecret
```

用到的每个 Secret 都需要在 spec.volumes 中指明，如果 Pod 中有多个容器，每个容器都需要自己的 volumeMounts 配置块，但是每个 Secret 只需要一个 spec.volumes，可以根据自己的应用场景将多个文件打包到一个 Secret 中，或者使用多个 Secret。

（2）自定义文件名挂载

挂载 Secret 时，可以使用 spec.volumes.secret.items 字段修改每个 key 的目标路径，即控制 Secret Key 在容器中的映射路径。

比如：

```yaml
apiVersion: v1
kind: Pod
metadata:
  name: mypod
spec:
  containers:
  - name: mypod
    image: redis
    volumeMounts:
    - name: foo
      mountPath: "/etc/foo"
      readOnly: true
  volumes:
  - name: foo
    secret:
      secretName: mysecret
      items:
      - key: username
        path: my-group/my-username
```

上述挂载方式，将 mysecret 中的 username 存储到了 /etc/foo/my-group/my-username 文件中，而不是 /etc/foo/username（不指定 items），由于 items 没有指定 password，因此 password 不会被挂载。如果使用了 spec.volumes.secret.items，只有在 items 中指定的 key 才会被挂载。

挂载的 Secret 在容器中作为文件，我们可以在 Pod 中查看挂载的文件内容：

```
$ ls /etc/foo/
username
password
$ cat /etc/foo/username
admin
$ cat /etc/foo/password
1f2d1e2e67df
```

（3）Secret 作为环境变量

Secret 可以作为环境变量使用，步骤如下：

① 创建一个 Secret 或者使用一个已存在的 Secret，多个 Pod 可以引用同一个 Secret。
② 为每个容器添加对应的 Secret Key 环境变量 env.valueFrom.secretKeyRef。

比如，定义 SECRET_USERNAME 和 SECRET_PASSWORD 两个环境变量，其值来自于名字为 mysecret 的 Secret：

```yaml
apiVersion: v1
kind: Pod
metadata:
  name: secret-env-pod
spec:
  containers:
  - name: mycontainer
    image: redis
    env:
      - name: SECRET_USERNAME
        valueFrom:
          secretKeyRef:
            name: mysecret
            key: username
      - name: SECRET_PASSWORD
        valueFrom:
          secretKeyRef:
            name: mysecret
            key: password
  restartPolicy: Never
```

挂载成功后，可以在容器中使用此变量：

```
$ echo $SECRET_USERNAME
admin
$ echo $SECRET_PASSWORD
1f2d1e2e67df
```

4. Secret 文件权限

Secret 默认挂载的文件的权限为 0644，可以通过 defaultMode 方式更改权限：

```yaml
apiVersion: v1
kind: Pod
metadata:
  name: mypod
spec:
  containers:
  - name: mypod
    image: redis
    volumeMounts:
    - name: foo
      mountPath: "/etc/foo"
  volumes:
  - name: foo
    secret:
```

```
        secretName: mysecret
        defaultMode: 256
```

更改的 Secret 挂载到/etc/foo 目录的文件权限为 0400。新版本可以直接指定 400。

5. imagePullSecret

在拉取私有镜像库中的镜像时，可能需要认证后才可拉取，此时可以使用 imagePullSecret 将包含 Docker 镜像注册表密码的 Secret 传递给 Kubelet，然后即可拉取私有镜像。

Kubernetes 支持在 Pod 中指定 Registry Key，用于拉取私有镜像仓库中的镜像。

首先创建一个镜像仓库账户信息的 Secret：

```
$kubectl create secret docker-registry myregistrykey
--docker-server=DOCKER_REGISTRY_SERVER --docker-username=DOCKER_USER
--docker-password=DOCKER_PASSWORD --docker-email=DOCKER_EMAIL
```

如果需要访问多个 Registry，则可以为每个注册表创建一个 Secret，在 Pods 拉取镜像时，Kubelet 会合并 imagePullSecrets 到.docker/config.json。注意 Secret 需要和 Pod 在同一个命名空间中。

创建完 imagePullSecrets 后，可以使用 imagePullSecrets 的方式引用该 Secret：

```
apiVersion: v1
kind: Pod
metadata:
  name: foo
  namespace: awesomeapps
spec:
  containers:
    - name: foo
      image: janedoe/awesomeapp:v1
  imagePullSecrets:
    - name: myregistrykey
```

6. 使用案例

本节演示的是 Secret 的一些常用配置，比如配置 SSH 密钥、创建隐藏文件等。

（1）定义包含 SSH 密钥的 Pod

首先，创建一个包含 SSH Key 的 Secret：

```
$kubectl create secret generic ssh-key-secret
--from-file=ssh-privatekey=/path/to/.ssh/id_rsa
--from-file=ssh-publickey=/path/to/.ssh/id_rsa.pub
```

然后将其挂载使用：

```
kind: Pod
apiVersion: v1
metadata:
  name: secret-test-pod
  labels:
    name: secret-test
spec:
  volumes:
  - name: secret-volume
    secret:
```

```
      secretName: ssh-key-secret
  containers:
  - name: ssh-test-container
    image: mySshImage
    volumeMounts:
    - name: secret-volume
      readOnly: true
      mountPath: "/etc/secret-volume"
```

上述密钥会被挂载到/etc/secret-volume。注意，挂载 SSH Key 需要考虑安全性的问题。

（2）创建隐藏文件

为了将数据"隐藏"起来（即文件名以句点符号开头的文件），可以让 Key 以一个句点符号开始，比如定义一个以句点符号开头的 Secret：

```
kind: Secret
apiVersion: v1
metadata:
  name: dotfile-secret
data:
  .secret-file: dmFsdWUtMg0KDQo=
```

挂载使用：

```
kind: Pod
apiVersion: v1
metadata:
  name: secret-dotfiles-pod
spec:
  volumes:
  - name: secret-volume
    secret:
      secretName: dotfile-secret
  containers:
  - name: dotfile-test-container
    image: K8S.gcr.io/busybox
    command:
    - ls
    - "-l"
    - "/etc/secret-volume"
    volumeMounts:
    - name: secret-volume
      readOnly: true
      mountPath: "/etc/secret-volume"
```

此时会在/etc/secret-volume 下创建一个.secret-file 的文件。

2.2.11　HPA

1. 什么是 HPA

HPA（Horizontal Pod Autoscaler，水平 Pod 自动伸缩器）可根据观察到的 CPU、内存使用率或自定义度量标准来自动扩展或缩容 Pod 的数量。HPA 不适用于无法缩放的对象，比如 DaemonSet。

HPA 控制器会定期调整 RC 或 Deployment 的副本数，以使观察到的平均 CPU 利用率与用户指定的目标相匹配。

HPA 需要 metrics-server（项目地址：https://github.com/kubernetes-incubator/metrics-server）获取度量指标，由于在高可用集群安装中已经安装了 metrics-server，所以本节的实践部分无须再次安装。

2. HPA 实践

在生产环境中，总会有一些意想不到的事情发生，比如公司网站流量突然升高，此时之前创建的 Pod 已不足以撑住所有的访问，而运维人员也不可能 24 小时守着业务服务，这时就可以通过配置 HPA，实现负载过高的情况下自动扩容 Pod 副本数以分摊高并发的流量，当流量恢复正常后，HPA 会自动缩减 Pod 的数量。

本节将测试实现一个 Web 服务器的自动伸缩特性，具体步骤如下：

首先启动一个 Nginx 服务：

```
[root@K8S-master01 ~]#kubectl run nginx-server --requests=cpu=10m --image=nginx --port=80
service/php-apache created
deployment.apps/php-apache created
```

临时开启 nginx-server 的端口，实际使用时需要定义 service：

```
kubectl expose deployment nginx-server --port=80
```

使用 kubectl autoscale 创建 HPA：

```
[root@K8S-master01 ~]# kubectl autoscale deployment nginx-server --cpu-percent=10 --min=1 --max=10
```

此 HPA 将根据 CPU 的使用率自动增加和减少副本数量，上述设置的是 CPU 使用率超过 10%（--cpu-percent 参数指定）即会增加 Pod 的数量，以保持所有 Pod 的平均 CPU 利用率为 10%，允许最大的 Pod 数量为 10（--max），最少的 Pod 数为 1（--min）。

查看当前 HPA 状态，因为未对其发送任何请求，所以当前 CPU 使用率为 0%：

```
[root@K8S-master01 metric-server]# kubectl get hpa
NAME           REFERENCE                TARGETS   MINPODS   MAXPODS   REPLICAS   AGE
nginx-server   Deployment/nginx-server  0%/10%    1         10        1          5m
```

查看当前 Nginx 的 Service 地址：

```
[root@K8S-master01 ~]# kubectl get service -n default
NAME           TYPE        CLUSTER-IP       EXTERNAL-IP   PORT(S)   AGE
kubernetes     ClusterIP   10.96.0.1        <none>        443/TCP   1d
nginx-server   ClusterIP   10.108.160.23    <none>        80/TCP    5m
```

增加负载：

```
[root@K8S-master01 ~]# while true; do wget -q -O- http://10.108.160.23 > /dev/null; done
```

1 分钟左右再次查看 HPA：

```
[root@K8S-master01 metric-server]# kubectl get hpa
NAME          REFERENCE                TARGETS    MINPODS   MAXPODS   REPLICAS   AGE
nginx-server  Deployment/nginx-server  540%/10%   1         10        1          15m
```

再次查看 Pod，可以看到 nginx-server 的 Pod 已经在扩容阶段：

```
[root@K8S-master01 metric-server]# kubectl get po
NAME                              READY   STATUS              RESTARTS   AGE
nginx-server-589c8db585-5cbxl     0/1     ContainerCreating   0          <invalid>
nginx-server-589c8db585-7whl8     1/1     Running             0          <invalid>
nginx-server-589c8db585-cv4hs     1/1     Running             0          <invalid>
nginx-server-589c8db585-m5dn6     0/1     ContainerCreating   0          <invalid>
nginx-server-589c8db585-sxbfm     1/1     Running             0          19m
nginx-server-589c8db585-xbctd     0/1     ContainerCreating   0          <invalid>
nginx-server-589c8db585-xffs9     1/1     Running             0          <invalid>
nginx-server-589c8db585-xlb8s     0/1     ContainerCreating   0          <invalid>
```

在增加负荷的终端，按 Ctrl+C 键终止访问。

停止 1 分钟后再次查看 HPA 和 deployment，此时副本已经恢复为 1：

```
[root@K8S-master01 metric-server]# kubectl get hpa
NAME          REFERENCE                TARGETS   MINPODS   MAXPODS   REPLICAS   AGE
nginx-server  Deployment/nginx-server  0%/10%    1         10        10         20m
```

2.2.12 Storage

本节介绍 Kubernetes Storage 的相关概念与使用，一般做持久化或者有状态的应用程序才会用到 Storage。

1. Volumes

Container（容器）中的磁盘文件是短暂的，当容器崩溃时，kubelet 会重新启动容器，但最初的文件将丢失，Container 会以最干净的状态启动。另外，当一个 Pod 运行多个 Container 时，各个容器可能需要共享一些文件。Kubernetes Volume 可以解决这两个问题。

（1）背景

Docker 也有卷的概念，但是在 Docker 中卷只是磁盘上或另一个 Container 中的目录，其生命周期不受管理。虽然目前 Docker 已经提供了卷驱动程序，但是功能非常有限，例如从 Docker 1.7 版本开始，每个 Container 只允许一个卷驱动程序，并且无法将参数传递给卷。

另一方面，Kubernetes 卷具有明确的生命周期，与使用它的 Pod 相同。因此，在 Kubernetes

中的卷可以比 Pod 中运行的任何 Container 都长,并且可以在 Container 重启或者销毁之后保留数据。Kubernetes 支持多种类型的卷,Pod 可以同时使用任意数量的卷。

从本质上讲,卷只是一个目录,可能包含一些数据,Pod 中的容器可以访问它。要使用卷 Pod 需要通过 .spec.volumes 字段指定为 Pod 提供的卷,以及使用 .spec.containers.volumeMounts 字段指定卷挂载的目录。从容器中的进程可以看到由 Docker 镜像和卷组成的文件系统视图,卷无法挂载其他卷或具有到其他卷的硬链接,Pod 中的每个 Container 必须独立指定每个卷的挂载位置。

(2) 卷的类型

Kubernetes 支持的卷的类型有很多,以下为常用的卷。

① awsElasticBlockStore(EBS)

awsElasticBlockStore 卷挂载一个 AWS EBS Volume 到 Pod 中,与 emptyDir 卷不同的是,当移除 Pod 时 EBS 卷的内容不会被删除,这意味着可以将数据预先放置在 EBS 卷中,并且可以在 Pod 之间切换该数据。

使用 awsElasticBlockStore 卷的限制:

- 运行 Pod 的节点必须是 AWS EC2 实例。
- AWS EC2 实例需要和 EBS 卷位于同一区域和可用区域。
- EBS 仅支持挂载卷的单个 EC2 实例。

在将 Pod 与 EBS 卷一起使用之前,需要先创建 EBS 卷,确保该卷的区域与集群的区域匹配,并检查 size 和 EBS 卷类型是否合理:

```
aws ec2 create-volume --availability-zone=eu-west-1a --size=10 --volume-type=gp2
```

AWS EBS 示例配置:

```
apiVersion: v1
kind: Pod
metadata:
  name: test-ebs
spec:
  containers:
  - image: K8S.gcr.io/test-webserver
    name: test-container
    volumeMounts:
    - mountPath: /test-ebs
      name: test-volume
  volumes:
  - name: test-volume
    # This AWS EBS volume must already exist.
    awsElasticBlockStore:
      volumeID: <volume-id>
      fsType: ext4
```

② CephFS

CephFS 卷允许将一个已经存在的卷挂载到 Pod 中,和 emptyDir 卷不同的是,当移除 Pod 时,CephFS 卷的内容不会被删除,这意味着可以将数据预先放置在 CephFS 卷中,并且可以在 Pod 之间切换该数据。CephFS 卷可以被多个写设备同时挂载。

和 AWS EBS 一样，需要先创建 CephFS 卷后才能使用它。

关于 CephFS 的更多内容，可以参考以下文档：

https://github.com/kubernetes/examples/tree/master/staging/volumes/cephfs/

③ConfigMap

ConfigMap 卷也可以作为 volume 使用，存储在 ConfigMap 中的数据可以通过 ConfigMap 类型的卷挂载到 Pod 中，然后使用该 ConfigMap 中的数据。引用 ConfigMap 对象时，只需要在 volume 中引用 ConfigMap 的名称即可，同时也可以自定义 ConfigMap 的挂载路径。

例如，将名称为 log-config 的 ConfigMap 挂载到 Pod 的/etc/config 目录下，挂载的文件名称为 path 指定的值，当前为 log_level：

```yaml
apiVersion: v1
kind: Pod
metadata:
  name: configmap-pod
spec:
  containers:
    - name: test
      image: busybox
      volumeMounts:
        - name: config-vol
          mountPath: /etc/config
  volumes:
    - name: config-vol
      configMap:
        name: log-config
        items:
          - key: log_level
            path: log_level
```

> **注 意**
>
> ConfigMap 需要提前创建。

④emptyDir

和上述 volume 不同的是，如果删除 Pod，emptyDir 卷中的数据也将被删除，一般 emptyDir 卷用于 Pod 中的不同 Container 共享数据。它可以被挂载到相同或不同的路径上。

默认情况下，emptyDir 卷支持节点上的任何介质，可能是 SSD、磁盘或网络存储，具体取决于自身的环境。可以将 emptyDir.medium 字段设置为 Memory，让 Kubernetes 使用 tmpfs（内存支持的文件系统），虽然 tmpfs 非常快，但是 tmpfs 在节点重启时，数据同样会被清除，并且设置的大小会被计入到 Container 的内存限制当中。

使用 emptyDir 卷的示例，直接指定 emptyDir 为{}即可：

```yaml
apiVersion: v1
kind: Pod
metadata:
  name: test-pd
spec:
```

```
containers:
- image: K8S.gcr.io/test-webserver
  name: test-container
  volumeMounts:
  - mountPath: /cache
    name: cache-volume
volumes:
- name: cache-volume
  emptyDir: {}
```

⑤GlusterFS

GlusterFS（以下简称为 GFS）是一个开源的网络文件系统，常被用于为 Kubernetes 提供动态存储，和 emptyDir 不同的是，删除 Pod 时 GFS 卷中的数据会被保留。

关于 GFS 的使用示例请参看 3.1 节。

⑥hostPath

hostPath 卷可将节点上的文件或目录挂载到 Pod 上，用于 Pod 自定义日志输出或访问 Docker 内部的容器等。

使用 hostPath 卷的示例。将主机的/data 目录挂载到 Pod 的/test-pd 目录：

```
apiVersion: v1
kind: Pod
metadata:
  name: test-pd
spec:
  containers:
  - image: K8S.gcr.io/test-webserver
    name: test-container
    volumeMounts:
    - mountPath: /test-pd
      name: test-volume
  volumes:
  - name: test-volume
    hostPath:
      # directory location on host
      path: /data
      # this field is optional
      type: Directory
```

hostPath 卷常用的 type（类型）如下。

- type 为空字符串：默认选项，意味着挂载 hostPath 卷之前不会执行任何检查。
- DirectoryOrCreate：如果给定的 path 不存在任何东西，那么将根据需要创建一个权限为 0755 的空目录，和 Kubelet 具有相同的组和权限。
- Directory：目录必须存在于给定的路径下。
- FileOrCreate：如果给定的路径不存储任何内容，则会根据需要创建一个空文件，权限设置为 0644，和 Kubelet 具有相同的组和所有权。
- File：文件，必须存在于给定路径中。
- Socket：UNIX 套接字，必须存在于给定路径中。
- CharDevice：字符设备，必须存在于给定路径中。

- BlockDevice：块设备，必须存在于给定路径中。

⑦NFS

NFS 卷也是一种网络文件系统，同时也可以作为动态存储，和 GFS 类似，删除 Pod 时，NFS 中的数据不会被删除。NFS 可以被多个写入同时挂载。

关于 NFS 的使用，请参考第 3 章。

⑧persistentVolumeClaim

persistentVolumeClaim 卷用于将 PersistentVolume（持久化卷）挂载到容器中，PersistentVolume 分为动态存储和静态存储，静态存储的 PersistentVolume 需要手动提前创建 PV，动态存储无需手动创建 PV。

⑨Secret

Secret 卷和 ConfigMap 卷类似，详情见 2.2.10 节。

⑩SubPath

有时可能需要将一个卷挂载到不同的子目录，此时使用 volumeMounts.subPath 可以实现不同子目录的挂载。

本示例为一个 LAMP 共享一个卷，使用 subPath 卷挂载不同的目录：

```
apiVersion: v1
kind: Pod
metadata:
  name: my-lamp-site
spec:
  containers:
  - name: mysql
    image: mysql
    env:
    - name: MYSQL_ROOT_PASSWORD
      value: "rootpasswd"
    volumeMounts:
    - mountPath: /var/lib/mysql
      name: site-data
      subPath: mysql
  - name: php
    image: php:7.0-apache
    volumeMounts:
    - mountPath: /var/www/html
      name: site-data
      subPath: html
  volumes:
  - name: site-data
    persistentVolumeClaim:
      claimName: my-lamp-site-data
```

更多 volume 可参考：

https://kubernetes.io/docs/concepts/storage/volumes/

2. PersistentVolume

管理计算资源需要关注的另一个问题是管理存储，PersistentVolume 子系统为用户和管理提供了一个 API，用于抽象如何根据使用类型提供存储的详细信息。为此，Kubernetes 引入了两个新的 API 资源：PersistentVolume 和 PersistentVolumeClaim。

PersistentVolume（简称 PV）是由管理员设置的存储，它同样是集群中的一类资源，PV 是容量插件，如 Volumes（卷），但其生命周期独立使用 PV 的任何 Pod，PV 的创建可使用 NFS、iSCSI、GFS、CEPH 等。

PersistentVolumeClaim（简称 PVC）是用户对存储的请求，类似于 Pod，Pod 消耗节点资源，PVC 消耗 PV 资源，Pod 可以请求特定级别的资源（CPU 和内存），PVC 可以请求特定的大小和访问模式。例如，可以以一次读/写或只读多次的模式挂载。

虽然 PVC 允许用户使用抽象存储资源，但是用户可能需要具有不同性质的 PV 来解决不同的问题，比如使用 SSD 硬盘来提高性能。所以集群管理员需要能够提供各种 PV，而不仅是大小和访问模式，并且无须让用户了解这些卷的实现方式，对于这些需求可以使用 StorageClass 资源实现。

目前 PV 的提供方式有两种：静态或动态。

静态 PV 由管理员提前创建，动态 PV 无需提前创建，只需指定 PVC 的 StorageClasse 即可。

（1）回收策略

当用户使用完卷时，可以从 API 中删除 PVC 对象，从而允许回收资源。回收策略会告诉 PV 如何处理该卷，目前卷可以保留、回收或删除。

- Retain：保留，该策略允许手动回收资源，当删除 PVC 时，PV 仍然存在，volume 被视为已释放，管理员可以手动回收卷。
- Recycle：回收，如果 volume 插件支持，Recycle 策略会对卷执行 rm -rf 清理该 PV，并使其可用于下一个新的 PVC，但是本策略已弃用，建议使用动态配置。
- Delete：删除，如果 volume 插件支持，删除 PVC 时会同时删除 PV，动态卷默认为 Delete。

（2）创建 PV

在使用持久化时，需要先创建 PV，然后再创建 PVC，PVC 会和匹配的 PV 进行绑定，然后 Pod 即可使用该存储。

创建一个基于 NFS 的 PV：

```
apiVersion: v1
kind: PersistentVolume
metadata:
  name: pv0003
spec:
  capacity:
    storage: 5Gi
  volumeMode: Filesystem
  accessModes:
    - ReadWriteOnce
  persistentVolumeReclaimPolicy: Recycle
  storageClassName: slow
  mountOptions:
    - hard
    - nfsvers=4.1
  nfs:
```

```
    path: /tmp
    server: 172.17.0.2
```

说明

- capacity：容量。
- accessModes：访问模式。包括以下 3 种：
  - ReadWriteOnce：可以被单节点以读写模式挂载，命令行中可以被缩写为 RWO。
  - ReadOnlyMany：可以被多个节点以只读模式挂载，命令行中可以被缩写为 ROX。
  - ReadWriteMany：可以被多个节点以读写模式挂载，命令行中可以被缩写为 RWX。
- storageClassName：PV 的类，一个特定类型的 PV 只能绑定到特定类别的 PVC。
- persistentVolumeReclaimPolicy：回收策略。
- mountOptions：非必须，新版本中已弃用。
- nfs：NFS 服务配置。包括以下两个选项：
  - path：NFS 上的目录
  - server：NFS 的 IP 地址

创建的 PV 会有以下几种状态：

- Available（可用），没有被 PVC 绑定的空间资源。
- Bound（已绑定），已经被 PVC 绑定。
- Released（已释放），PVC 被删除，但是资源还未被重新使用。
- Failed（失败），自动回收失败。

可以创建一个基于 hostPath 的 PV：

```
kind: PersistentVolume
apiVersion: v1
metadata:
  name: task-pv-volume
  labels:
    type: local
spec:
  storageClassName: manual
  capacity:
    storage: 10Gi
  accessModes:
    - ReadWriteOnce
  hostPath:
    path: "/mnt/data"
```

（3）创建 PVC

创建 PVC 需要注意的是，各个方面都符合要求 PVC 才能和 PV 进行绑定，比如 accessModes、storageClassName、volumeMode 都需要相同才能进行绑定。

创建 PVC 的示例如下：

```
kind: PersistentVolumeClaim
apiVersion: v1
metadata:
```

```yaml
  name: myclaim
spec:
  accessModes:
    - ReadWriteOnce
  volumeMode: Filesystem
  resources:
    requests:
      storage: 8Gi
  storageClassName: slow
  selector:
    matchLabels:
      release: "stable"
    matchExpressions:
      - {key: environment, operator: In, values: [dev]}
```

比如上述基于 hostPath 的 PV 可以使用以下 PVC 进行绑定，storage 可以比 PV 小：

```yaml
kind: PersistentVolumeClaim
apiVersion: v1
metadata:
  name: task-pv-claim
spec:
  storageClassName: manual
  accessModes:
    - ReadWriteOnce
  resources:
    requests:
      storage: 3Gi
```

然后创建一个 Pod 指定 volumes 即可使用这个 PV：

```yaml
kind: Pod
apiVersion: v1
metadata:
  name: task-pv-pod
spec:
  volumes:
    - name: task-pv-storage
      persistentVolumeClaim:
        claimName: task-pv-claim
  containers:
    - name: task-pv-container
      image: nginx
      ports:
        - containerPort: 80
          name: "http-server"
      volumeMounts:
        - mountPath: "/usr/share/nginx/html"
          name: task-pv-storage
```

> **注 意**
>
> claimName 需要和上述定义的 PVC 名称 task-pv-claim 一致。

3. StorageClass

StorageClass 为管理员提供了一种描述存储"类"的方法,可以满足用户不同的服务质量级别、备份策略和任意策略要求的存储需求,一般动态 PV 都会通过 StorageClass 来定义。

每个 StorageClass 包含字段 provisioner、parameters 和 reclaimPolicy,StorageClass 对象的名称很重要,管理员在首次创建 StorageClass 对象时设置的类的名称和其他参数,在被创建对象后无法再更新这些对象。

定义一个 StorageClass 的示例如下:

```
kind: StorageClass
apiVersion: storage.K8S.io/v1
metadata:
  name: standard
provisioner: kubernetes.io/aws-ebs
parameters:
  type: gp2
reclaimPolicy: Retain
mountOptions:
  - debug
volumeBindingMode: Immediate
```

(1) Provisioner

StorageClass 有一个 provisioner 字段,用于指定配置 PV 的卷的类型,必须指定此字段,目前支持的卷插件如表 2-7 所示。

表 2-7 卷插件的类型

Volume Plugin	Internal Provisioner	Config Example
AWSElasticBlockStore	✓	AWS EBS
AzureFile	✓	Azure File
AzureDisk	✓	Azure Disk
CephFS	-	-
Cinder	✓	OpenStack Cinder
FC	-	-
Flexvolume	-	-
Flocker	✓	-
GCEPersistentDisk	✓	GCE PD
Glusterfs	✓	Glusterfs
iSCSI	-	-
Quobyte	✓	Quobyte
NFS	-	-
RBD	✓	Ceph RBD
VsphereVolume	✓	vSphere
PortworxVolume	✓	Portworx Volume
ScaleIO	✓	ScaleIO
StorageOS	✓	StorageOS
Local	-	Local

> **注 意**
>
> provisioner 不仅限于此处列出的内部 provisioner，还可以运行和指定外部供应商。例如，NFS 不提供内部配置程序，但是可以使用外部配置程序，外部配置方式参见以下网址：
> https://github.com/kubernetes-incubator/external-storage

（2）ReclaimPolicy

回收策略，可以是 Delete、Retain，默认为 Delete。

（3）MountOptions

通过 StorageClass 动态创建的 PV 可以使用 MountOptions 指定挂载参数。如果指定的卷插件不支持指定的挂载选项，就不会被创建成功，因此在设置时需要进行确认。

（4）Parameters

PVC 具有描述属于 StorageClass 卷的参数，根据具体情况，取决于 provisioner，可以接受不同类型的参数。比如，type 为 io1 和特定参数 iopsPerGB 是 EBS 所具有的。如果省略配置参数，将采用默认值。

4. 定义 StorageClass

StorageClass 一般用于定义动态存储卷，只需要在 Pod 上指定 StorageClass 的名字即可自动创建对应的 PV，无须再手工创建。

以下为常用的 StorageClass 定义方式。

（1）AWS EBS

```
kind: StorageClass
apiVersion: storage.K8S.io/v1
metadata:
  name: slow
provisioner: kubernetes.io/aws-ebs
parameters:
  type: io1
  iopsPerGB: "10"
  fsType: ext4
```

说明

- type: io1、gp2、sc1、st1，默认为 gp2。详情可查看以下网址的内容：

    https://docs.aws.amazon.com/AWSEC2/latest/UserGuide/EBSVolumeTypes.html

- iopsPerGB: 仅适用于 io1 卷，即每 GiB 每秒的 I/O 操作。
- fsType: Kubernetes 支持的 fsType，默认值为：ext4。

（2）GCE PD

```
kind: StorageClass
apiVersion: storage.K8S.io/v1
metadata:
  name: slow
```

```
provisioner: kubernetes.io/gce-pd
parameters:
  type: pd-standard
  replication-type: none
```

说明

- type: pd-standard 或 pd-ssd,默认为 pd-standard。
- replication-type: none 或 regional-pd,默认值为 none。

(3) GFS

```
apiVersion: storage.K8S.io/v1
kind: StorageClass
metadata:
  name: slow
provisioner: kubernetes.io/glusterfs
parameters:
  resturl: "http://127.0.0.1:8081"
  clusterid: "630372ccdc720a92c681fb928f27b53f"
  restauthenabled: "true"
  restuser: "admin"
  secretNamespace: "default"
  secretName: "heketi-secret"
  gidMin: "40000"
  gidMax: "50000"
  volumetype: "replicate:3"
```

说明

- resturl: Gluster REST 服务/Heketi 服务的 URL,这是 GFS 动态存储必需的参数。
- restauthenabled: 用于对 REST 服务器进行身份验证,此选项已被启用。如果需要启用身份验证,只需指定 restuser、restuserkey、secretName 或 secretNamespace 其中一个即可。
- restuser: 访问 Gluster REST 服务的用户。
- secretNamespace,secretName: 与 Gluster REST 服务交互时使用的 Secret。这些参数是可选的,如果没有身份认证不用配置此参数。该 Secret 使用 type 为 kubernetes.io/glusterfs 的 Secret 进行创建,例如:

```
kubectl create secret generic heketi-secret \
  --type="kubernetes.io/glusterfs" --from-literal=key='opensesame' \
  --namespace=default
```

- clusterid: Heketi 创建集群的 ID,可以是一个列表,用逗号分隔。
- gidMin,gidMax: StorageClass 的 GID 范围,可选,默认为 2000-2147483647。
- volumetype: 创建的 GFS 卷的类型,主要分为以下 3 种:
  - ➢ Replica 卷: volumetype: replicate:3,表示每个 PV 会创建 3 个副本。
  - ➢ Disperse/EC 卷: volumetype: disperse:4:2,其中 4 是数据,2 是冗余。
  - ➢ Distribute 卷: volumetype: none。

当使用 GFS 作为动态配置 PV 时,会自动创建一个格式为 gluster-dynamic-<claimname>的 Endpoint 和 Headless Service,删除 PVC 会自动删除 PV、Endpoint 和 Headless Service。

（4）Ceph RBD

```
kind: StorageClass
apiVersion: storage.K8S.io/v1
metadata:
  name: fast
provisioner: kubernetes.io/rbd
parameters:
  monitors: 10.16.153.105:6789
  adminId: kube
  adminSecretName: ceph-secret
  adminSecretNamespace: kube-system
  pool: kube
  userId: kube
  userSecretName: ceph-secret-user
  userSecretNamespace: default
  fsType: ext4
  imageFormat: "2"
  imageFeatures: "layering"
```

说明

- monitors：Ceph 的 monitor，用逗号分隔，此参数是必需的。
- adminId：Ceph 客户端的 ID，默认为 admin。
- adminSecretName：adminId 的 Secret 名称，此参数是必需的，该 Secret 必须是 kubernetes.io/rbd 类型。
- adminSecretNamespace：Secret 所在的 NameSpace（命名空间），默认为 default。
- pool：Ceph RBD 池，默认为 rbd。
- userId：Ceph 客户端 ID，默认值与 adminId 相同。
- userSecretName：和 adminSecretName 类似，必须与 PVC 存在于同一个命名空间，创建方式如下：

  ```
  kubectl create secret generic ceph-secret --type="kubernetes.io/rbd" \
  --from-literal=key='QVFEQ1pMdFhPUnQrSmhBQUFYaERWNHJsZ3BsMmNjcDR6RFZST0E9PQ==' \
     --namespace=kube-system
  ```

- imageFormat：Ceph RBD 镜像格式，默认值为 2，旧一些的为 1。
- imagefeatures：可选参数，只有设置 imageFormat 为 2 时才能使用，目前仅支持 layering。

更多详情请参考以下网址：

https://kubernetes.io/docs/concepts/storage/storage-classes/

4. 动态存储卷

动态卷的配置允许按需自动创建 PV，如果没有动态配置，集群管理员必须手动创建 PV。
动态卷的配置基于 StorageClass API 组中的 API 对象 storage.K8S.io。

（1）定义 GCE 动态预配置

要启用动态配置，集群管理员需要为用户预先创建一个或多个 StorageClass 对象，比如创建一个名字为 slow 且使用 gce 提供存储卷的 StorageClass：

```
apiVersion: storage.K8S.io/v1
kind: StorageClass
metadata:
  name: slow
provisioner: kubernetes.io/gce-pd
parameters:
  type: pd-standard
```

再例如创建一个能提供 SSD 磁盘的 StorageClass：

```
apiVersion: storage.K8S.io/v1
kind: StorageClass
metadata:
  name: fast
provisioner: kubernetes.io/gce-pd
parameters:
  type: pd-ssd
```

用户通过定义包含 StorageClass 的 PVC 来请求动态调配的存储。在 Kubernetesv 1.6 之前，是通过 volume.beta.kubernetes.io/storage-class 注解来完成的。在 1.6 版本之后，此注解已弃用。

例如，创建一个快速存储类，定义的 PersistentVolumeClaim 如下：

```
apiVersion: v1
kind: PersistentVolumeClaim
metadata:
  name: claim1
spec:
  accessModes:
    - ReadWriteOnce
  storageClassName: fast
  resources:
    requests:
      storage: 30Gi
```

> **注 意**
>
> storageClassName 要与上述创建的 StorageClass 名字相同。

之后会自动创建一个 PV 与该 PVC 进行绑定，然后 Pod 即可挂载使用。

（2）定义 GFS 动态预配置

可以参考 3.1 节定义一个 GFS 的 StorageClass：

```
apiVersion: storage.K8S.io/v1
kind: StorageClass
metadata:
  name: gluster-heketi
provisioner: kubernetes.io/glusterfs
parameters:
  resturl: "http://10.111.95.240:8080"
  restauthenabled: "false"
```

之后定义一个 PVC：

```yaml
kind: PersistentVolumeClaim
apiVersion: v1
metadata:
  name: pvc-gluster-heketi
spec:
  accessModes: [ "ReadWriteOnce" ]
  storageClassName: "gluster-heketi"
  resources:
    requests:
      storage: 1Gi
```

PVC 一旦被定义，系统便发出 Heketi 进行相应的操作，在 GFS 集群上创建 brick，再创建并启动一个 volume。

然后定义一个 Pod 使用该存储卷：

```yaml
apiVersion: v1
kind: Pod
metadata:
  name: pod-use-pvc
spec:
  containers:
  - name: pod-use-pvc
    image: busybox
    command:
    - sleep
    - "3600"
    volumeMounts:
    - name: gluster-volume
      mountPath: "/pv-data"
      readOnly: false
  volumes:
  - name: gluster-volume
    persistentVolumeClaim:
      claimName: pvc-gluster-heketi
```

claimName 为上述创建的 PVC 的名称。

2.2.13 Service

Service 主要用于 Pod 之间的通信，对于 Pod 的 IP 地址而言，Service 是提前定义好并且是不变的资源类型。

1. 基本概念

Kubernetes Pod 具有生命周期的概念，它可以被创建、删除、销毁，一旦被销毁就意味着生命周期的结束。通过 ReplicaSet 能够动态地创建和销毁 Pod，例如进行扩缩容和执行滚动升级。每个 Pod 都会获取到它自己的 IP 地址，但是这些 IP 地址不总是稳定和可依赖的，这样就会导致一个问题：在 Kubernetes 集群中，如果一组 Pod（比如后端的 Pod）为其他 Pod（比如前端的 Pod）提供服务，那么如果它们之间使用 Pod 的 IP 地址进行通信，在 Pod 重建后，将无法再进行连接。

为了解决上述问题，Kubernetes 引用了 Service 这样一种抽象概念：逻辑上的一组 Pod，即一种可以访问 Pod 的策略——通常称为微服务。这一组 Pod 能够被 Service 访问到，通常是通过 Label Selector（标签选择器）实现的。

举个例子，有一个用作图片处理的 backend（后端），运行了 3 个副本，这些副本是可互换的，所以 frontend（前端）不需要关心它们调用了哪个 backend 副本，然而组成这一组 backend 程序的 Pod 实际上可能会发生变化，即便这样 frontend 也没有必要知道，而且也不需要跟踪这一组 backend 的状态，因为 Service 能够解耦这种关联。

对于 Kubernetes 集群中的应用，Kubernetes 提供了简单的 Endpoints API，只要 Service 中的一组 Pod 发生变更，应用程序就会被更新。对非 Kubernetes 集群中的应用，Kubernetes 提供了基于 VIP 的网桥的方式访问 Service，再由 Service 重定向到 backend Pod。

2. 定义 Service

一个 Service 在 Kubernetes 中是一个 REST 对象，和 Pod 类似。像所有 REST 对象一样，Service 的定义可以基于 POST 方式，请求 APIServer 创建新的实例。例如，假定有一组 Pod，它们暴露了 9376 端口，同时具有 app=MyApp 标签。此时可以定义 Service 如下：

```
kind: Service
apiVersion: v1
metadata:
  name: my-service
spec:
  selector:
    app: MyApp
  ports:
    - protocol: TCP
      port: 80
      targetPort: 9376
```

上述配置创建一个名为 my-service 的 Service 对象，它会将请求代理到 TCP 端口为 9376 并且具有标签 app=MyApp 的 Pod 上。这个 Service 会被分配一个 IP 地址，通常称为 ClusterIP，它会被服务的代理使用。

需要注意的是，Service 能够将一个接收端口映射到任意的 targetPort。默认情况下，targetPort 将被设置为与 Port 字段相同的值。targetPort 可以设置为一个字符串，引用 backend Pod 的一个端口的名称。

Kubernetes Service 能够支持 TCP 和 UDP 协议，默认为 TCP 协议。

3. 定义没有 Selector 的 Service

Service 抽象了该如何访问 Kubernetes Pod，但也能够抽象其他类型的 backend，例如：

- 希望在生产环境中访问外部的数据库集群。
- 希望 Service 指向另一个 NameSpace 中或其他集群中的服务。
- 正在将工作负载转移到 Kubernetes 集群，和运行在 Kubernetes 集群之外的 backend。

在任何这些场景中，都能定义没有 Selector 的 Service：

```
kind: Service
apiVersion: v1
metadata:
```

```yaml
  name: my-service
spec:
  ports:
    - protocol: TCP
      port: 80
      targetPort: 9376
```

由于这个 Service 没有 Selector，就不会创建相关的 Endpoints 对象，可以手动将 Service 映射到指定的 Endpoints：

```yaml
kind: Endpoints
apiVersion: v1
metadata:
  name: my-service
subsets:
  - addresses:
      - ip: 1.2.3.4
    ports:
      - port: 9376
```

> **注　意**
>
> Endpoint IP 地址不能是 loopback（127.0.0.0/8）、link-local（169.254.0.0/16）或者 link-local 多播地址（224.0.0.0/24）。

访问没有 Selector 的 Service 与有 Selector 的 Service 的原理相同。请求将被路由到用户定义的 Endpoint，该示例为 1.2.3.4:9376。

ExternalName Service 是 Service 的特例，它没有 Selector，也没有定义任何端口和 Endpoint，它通过返回该外部服务的别名来提供服务。

比如当查询主机 my-service.prod.svc 时，集群的 DNS 服务将返回一个值为 my.database.example.com 的 CNAME 记录：

```yaml
kind: Service
apiVersion: v1
metadata:
  name: my-service
  namespace: prod
spec:
  type: ExternalName
  externalName: my.database.example.com
```

4. VIP 和 Service 代理

在 Kubernetes 集群中，每个节点运行一个 kube-proxy 进程。kube-proxy 负责为 Service 实现了一种 VIP（虚拟 IP）的形式，而不是 ExternalName 的形式。在 Kubernetesv 1.0 版本中，代理完全是 userspace。在 Kubernetesv 1.1 版中新增了 iptables 代理，从 Kubernetesv 1.2 版起，默认是 iptables 代理。从 Kubernetesv 1.8 版开始新增了 ipvs 代理，生产环境建议使用 ipvs 模式。

在 Kubernetesv 1.0 版中 Service 是 4 层（TCP/UDP over IP）概念，在 Kubernetesv 1.1 版中新增了 Ingress API（beta 版），用来表示 7 层（HTTP）服务。

（1）iptables 代理模式

这种模式下 kube-proxy 会监视 Kubernetes Master 对 Service 对象和 Endpoints 对象的添加和移除。对每个 Service 它会创建 iptables 规则，从而捕获到该 Service 的 ClusterIP（虚拟 IP）和端口的请求，进而将请求重定向到 Service 的一组 backend 中的某个 Pod 上面。对于每个 Endpoints 对象，它也会创建 iptables 规则，这个规则会选择一个 backend Pod。

默认的策略是随机选择一个 backend，如果要实现基于客户端 IP 的会话亲和性，可以将 service.spec.sessionAffinity 的值设置为 ClusterIP（默认为 None）。

和 userspace 代理类似，网络返回的结果都是到达 Service 的 IP:Port 请求，这些请求会被代理到一个合适的 backend，不需要客户端知道关于 Kubernetes、Service 或 Pod 的任何信息。这比 userspace 代理更快、更可靠，并且当初始选择的 Pod 没有响应时，iptables 代理能够自动重试另一个 Pod。

（2）ipvs 代理模式

在此模式下，kube-proxy 监视 Kubernetes Service 和 Endpoint，调用 netlink 接口以相应地创建 ipvs 规则，并定期与 Kubernetes Service 和 Endpoint 同步 ipvs 规则，以确保 ipvs 状态与期望保持一致。访问服务时，流量将被重定向到其中一个后端 Pod。

与 iptables 类似，ipvs 基于 netfilter 钩子函数，但是 ipvs 使用哈希表作为底层数据结构并在内核空间中工作，这意味着 ipvs 可以更快地重定向流量，并且在同步代理规则时具有更好的性能，此外，ipvs 为负载均衡算法提供了更多的选项，例如：

- rr 轮询
- lc 最少连接
- dh 目标哈希
- sh 源哈希
- sed 预计延迟最短
- nq 从不排队

5. 多端口 Service

在许多情况下，Service 可能需要暴露多个端口，对于这种情况 Kubernetes 支持 Service 定义多个端口，但使用多个端口时，必须提供所有端口的名称，例如：

```
kind: Service
apiVersion: v1
metadata:
  name: my-service
spec:
  selector:
    app: MyApp
  ports:
  - name: http
    protocol: TCP
    port: 80
    targetPort: 9376
  - name: https
    protocol: TCP
```

```
    port: 443
    targetPort: 9377
```

6. 发布服务/服务类型

对于应用程序的某些部分（例如前端），一般要将服务公开到集群外部供用户访问。这种情况下都是用 Ingress 通过域名进行访问。

Kubernetes ServiceType（服务类型）主要包括以下几种：

- ClusterIP　在集群内部使用，默认值，只能从集群中访问。
- NodePort　在所有节点上打开一个端口，此端口可以代理至后端 Pod，可以通过 NodePort 从集群外部访问集群内的服务，格式为 NodeIP:NodePort。
- LoadBalancer　使用云提供商的负载均衡器公开服务，成本较高。
- ExternalName　通过返回定义的 CNAME 别名，没有设置任何类型的代理，需要 1.7 或更高版本 kube-dns 支持。

以 NodePort 为例。如果将 type 字段设置为 NodePort，则 Kubernetes 将从 --service-node-port-range 参数指定的范围（默认为 30000-32767）中自动分配端口，也可以手动指定 NodePort，并且每个节点将代理该端口到 Service。

一般格式如下：

```
kind: Service
apiVersion: v1
metadata:
  labels:
    K8S-app: kubernetes-dashboard
  name: kubernetes-dashboard
  namespace: kube-system
spec:
  type: NodePort
  ports:
    - port: 443
      targetPort: 8443
      nodePort: 30000
  selector:
    K8S-app: kubernetes-dashboard
```

常用的服务访问是 NodePort 和 Ingress（关于 Ingress 参看 2.2.14 节），其他服务访问方式详情参看以下网址：

https://kubernetes.io/docs/concepts/services-networking/service/#publishing-services-service-types

2.2.14　Ingress

Ingress 为 Kubernetes 集群中的服务提供了入口，可以提供负载均衡、SSL 终止和基于名称的虚拟主机，在生产环境中常用的 Ingress 有 Treafik、Nginx、HAProxy、Istio 等。

1. 基本概念

在 Kubernetesv 1.1 版中添加的 Ingress 用于从集群外部到集群内部 Service 的 HTTP 和 HTTPS

路由，流量从 Internet 到 Ingress 再到 Services 最后到 Pod 上，通常情况下，Ingress 部署在所有的 Node 节点上。

Ingress 可以配置提供服务外部访问的 URL、负载均衡、终止 SSL，并提供基于域名的虚拟主机。但 Ingress 不会暴露任意端口或协议。

2. 创建一个 Ingress

创建一个简单的 Ingress 如下：

```yaml
apiVersion: extensions/v1beta1
kind: Ingress
metadata:
  name: simple-fanout-example
  annotations:
    nginx.ingress.kubernetes.io/rewrite-target: /
spec:
  rules:
  - host: foo.bar.com
    http:
      paths:
      - path: /foo
        backend:
          serviceName: service1
          servicePort: 4200
      - path: /bar
        backend:
          serviceName: service2
          servicePort: 8080
```

上述 host 定义该 Ingress 的域名，将其解析至任意 Node 上即可访问。

如果访问的是 foo.bar.com/foo，则被转发到 service1 的 4200 端口。

如果访问的是 foo.bar.com/bar，则被转发到 service2 的 8080 端口。

（1）Ingress Rules
- host：可选，一般都会配置对应的域名。
- path：每个路径都有一个对应的 serviceName 和 servicePort，在流量到达服务之前，主机和路径都会与传入请求的内容匹配。
- backend：描述 Service 和 Port 的组合。对 Ingress 匹配主机和路径的 HTTP 与 HTTPS 请求将被发送到对应的后端。

（2）默认后端

没有匹配到任何规则的流量将被发送到默认后端。默认后端通常是 Ingress Controller 的配置选项，并未在 Ingress 资源中指定。

3. Ingress 类型

（1）单域名

单个域名匹配多个 path 到不同的 service：

```yaml
apiVersion: extensions/v1beta1
kind: Ingress
```

```yaml
metadata:
  name: simple-fanout-example
  annotations:
    nginx.ingress.kubernetes.io/rewrite-target: /
spec:
  rules:
  - host: foo.bar.com
    http:
      paths:
      - path: /foo
        backend:
          serviceName: service1
          servicePort: 4200
      - path: /bar
        backend:
          serviceName: service2
          servicePort: 8080
```

此时，访问 foo.bar.com/foo 到 service1 的 4200。访问 foo.bar.com/bar 到 service2 的 8080。

（2）多域名

基于域名的虚拟主机支持将 HTTP 流量路由到同一 IP 地址的多个主机名：

```yaml
apiVersion: extensions/v1beta1
kind: Ingress
metadata:
  name: name-virtual-host-ingress
spec:
  rules:
  - host: foo.bar.com
    http:
      paths:
      - backend:
          serviceName: service1
          servicePort: 80
  - host: bar.foo.com
    http:
      paths:
      - backend:
          serviceName: service2
          servicePort: 80
```

此时，访问 foo.bar.com 到 service1，访问 bar.foo.com 到 service2。

（3）基于 TLS 的 Ingress

首先创建证书，生产环境的证书为公司购买的证书：

```
kubectl -n default create secret tls nginx-test-tls --key=tls.key --cert=tls.crt
```

定义 Ingress（此示例为 Traefik，nginx-ingress 将 traefik 改为 nginx 即可）：

```yaml
apiVersion: extensions/v1beta1
kind: Ingress
metadata:
  name: nginx-https-test
```

```yaml
  namespace: default
  annotations:
    kubernetes.io/ingress.class: traefik
spec:
  rules:
  - host: traefix-test.com
    http:
      paths:
      - backend:
          serviceName: nginx-svc
          servicePort: 80
  tls:
  - secretName: nginx-test-tls
```

4. 更新 Ingress

更新 Ingress 可以直接使用 kubectl edit ingress INGRESS-NAME 进行更改，也可以通过 kubectlapply-f NEW-INGRESS-YAML.yaml 进行更改。

更多 Ingress 配置请参考第 5 章 Nginx Ingress 的内容。

2.2.15 Taint 和 Toleration

Taint 能够使节点排斥一类特定的 Pod，Taint 和 Toleration 相互配合可以用来避免 Pod 被分配到不合适的节点，比如 Master 节点不允许部署系统组件之外的其他 Pod。每个节点上都可以应用一个或多个 Taint，这表示对于那些不能容忍这些 Taint 的 Pod 是不会被该节点接受的。如果将 Toleration 应用于 Pod 上，则表示这些 Pod 可以（但不要求）被调度到具有匹配 Taint 的节点上。

1. 概念

给节点增加一个 Taint：

```
[root@K8S-master01 2.2.8]# kubectl taint nodes K8S-node01 key=value:NoSchedule
node/K8S-node01 tainted
```

上述命令给 K8S-node01 增加一个 Taint，它的 key 对应的就是键，value 对应就是值，effect 对应的就是 NoSchedule。这表明只有和这个 Taint 相匹配的 Toleration 的 Pod 才能够被分配到 K8S-node01 节点上。按如下方式在 PodSpec 中定义 Pod 的 Toleration，就可以将 Pod 部署到该节点上。

方式一：

```yaml
tolerations:
- key: "key"
  operator: "Equal"
  value: "value"
  effect: "NoSchedule"
```

方式二：

```yaml
tolerations:
- key: "key"
  operator: "Exists"
```

```
    effect: "NoSchedule"
```

一个 Toleration 和一个 Taint 相匹配是指它们有一样的 key 和 effect，并且如果 operator 是 Exists（此时 toleration 不指定 value）或者 operator 是 Equal，则它们的 value 应该相等。

注意两种情况：

- 如果一个 Toleration 的 key 为空且 operator 为 Exists，表示这个 Toleration 与任意的 key、value 和 effect 都匹配，即这个 Toleration 能容忍任意的 Taint：

```
tolerations:
- operator: "Exists"
```

- 如果一个 Toleration 的 effect 为空，则 key 与之相同的相匹配的 Taint 的 effect 可以是任意值：

```
tolerations:
- key: "key"
  operator: "Exists"
```

上述例子使用到 effect 的一个值 NoSchedule，也可以使用 PreferNoSchedule，该值定义尽量避免将 Pod 调度到存在其不能容忍的 Taint 的节点上，但并不是强制的。effect 的值还可以设置为 NoExecute。

一个节点可以设置多个 Taint，也可以给一个 Pod 添加多个 Toleration。Kubernetes 处理多个 Taint 和 Toleration 的过程就像一个过滤器：从一个节点的所有 Taint 开始遍历，过滤掉那些 Pod 中存在与之相匹配的 Toleration 的 Taint。余下未被过滤的 Taint 的 effect 值决定了 Pod 是否会被分配到该节点，特别是以下情况：

- 如果未被过滤的 Taint 中存在一个以上 effect 值为 NoSchedule 的 Taint，则 Kubernetes 不会将 Pod 分配到该节点。
- 如果未被过滤的 Taint 中不存在 effect 值为 NoExecute 的 Taint，但是存在 effect 值为 PreferNoSchedule 的 Taint，则 Kubernetes 会尝试将 Pod 分配到该节点。
- 如果未被过滤的 Taint 中存在一个以上 effect 值为 NoExecute 的 Taint，则 Kubernetes 不会将 Pod 分配到该节点（如果 Pod 还未在节点上运行），或者将 Pod 从该节点驱逐（如果 Pod 已经在节点上运行）。

例如，假设给一个节点添加了以下的 Taint：

```
kubectl taint nodes K8S-node01 key1=value1:NoSchedule
kubectl taint nodes K8S-node01 key1=value1:NoExecute
kubectl taint nodes K8S-node01 key2=value2:NoSchedule
```

然后存在一个 Pod，它有两个 Toleration：

```
tolerations:
- key: "key1"
  operator: "Equal"
  value: "value1"
  effect: "NoSchedule"
- key: "key1"
  operator: "Equal"
```

```
    value: "value1"
    effect: "NoExecute"
```

在上述例子中，该 Pod 不会被分配到上述节点，因为没有匹配第三个 Taint。但是如果给节点添加上述 3 个 Taint 之前，该 Pod 已经在上述节点中运行，那么它不会被驱逐，还会继续运行在这个节点上，因为第 3 个 Taint 是唯一不能被这个 Pod 容忍的。

通常情况下，如果给一个节点添加了一个 effect 值为 NoExecute 的 Taint，则任何不能容忍这个 Taint 的 Pod 都会马上被驱逐，任何可以容忍这个 Taint 的 Pod 都不会被驱逐。但是，如果 Pod 存在一个 effect 值为 NoExecute 的 Toleration 指定了可选属性 tolerationSeconds 的值，则该值表示是在给节点添加了上述 Taint 之后 Pod 还能继续在该节点上运行的时间，例如：

```
tolerations:
- key: "key1"
  operator: "Equal"
  value: "value1"
  effect: "NoExecute"
  tolerationSeconds: 3600
```

表示如果这个 Pod 正在运行，然后一个匹配的 Taint 被添加到其所在的节点，那么 Pod 还将继续在节点上运行 3600 秒，然后被驱逐。如果在此之前上述 Taint 被删除了，则 Pod 不会被驱逐。

删除一个 Taint：

```
kubectl taint nodes K8S-node01 key1:NoExecute-
```

查看 Taint：

```
[root@K8S-master01 2.2.8]# kubectl describe node K8S-node01 | grep Taint
Taints:             key=value:NoSchedule
```

2. 用例

通过 Taint 和 Toleration 可以灵活地让 Pod 避开某些节点或者将 Pod 从某些节点被驱逐。下面是几种情况。

（1）专用节点

如果想将某些节点专门分配给特定的一组用户使用，可以给这些节点添加一个 Taint（kubectl taint nodes nodename dedicated=groupName:NoSchedule），然后给这组用户的 Pod 添加一个相对应的 Toleration。拥有上述 Toleration 的 Pod 就能够被分配到上述专用节点，同时也能够被分配到集群中的其他节点。如果只希望这些 Pod 只能分配到上述专用节点中，那么还需要给这些专用节点另外添加一个和上述 Taint 类似的 Label（例如：dedicated=groupName），然后给 Pod 增加节点亲和性要求或者使用 NodeSelector，就能将 Pod 只分配到添加了 dedicated=groupName 标签的节点上。

（2）特殊硬件的节点

在部分节点上配备了特殊硬件（比如 GPU）的集群中，我们只允许特定的 Pod 才能部署在这些节点上。这时可以使用 Taint 进行控制，添加 Taint 如 kubectl taint nodes nodename special=true:NoSchedule 或者 kubectl taint nodes nodename special=true:PreferNoSchedule，然后给需要部署在这些节点上的 Pod 添加相匹配的 Toleration 即可。

（3）基于 Taint 的驱逐

属于 alpha 特性，在每个 Pod 中配置在节点出现问题时的驱逐行为。

3. 基于 Taint 的驱逐

之前提到过 Taint 的 effect 值 NoExecute，它会影响已经在节点上运行的 Pod。如果 Pod 不能忍受 effect 值为 NoExecute 的 Taint，那么 Pod 将会被马上驱逐。如果能够忍受 effect 值为 NoExecute 的 Taint，但是在 Toleration 定义中没有指定 tolerationSeconds，则 Pod 还会一直在这个节点上运行。

在 Kubernetes 1.6 版以后已经支持（alpha）当某种条件为真时，Node Controller 会自动给节点添加一个 Taint，用以表示节点的问题。当前内置的 Taint 包括：

- node.kubernetes.io/not-ready　节点未准备好，相当于节点状态 Ready 的值为 False。
- node.kubernetes.io/unreachable　Node Controller 访问不到节点，相当于节点状态 Ready 的值为 Unknown。
- node.kubernetes.io/out-of-disk　节点磁盘耗尽。
- node.kubernetes.io/memory-pressure　节点存在内存压力。
- node.kubernetes.io/disk-pressure　节点存在磁盘压力。
- node.kubernetes.io/network-unavailable　节点网络不可达。
- node.kubernetes.io/unschedulable　节点不可调度。
- node.cloudprovider.kubernetes.io/uninitialized　如果 Kubelet 启动时指定了一个外部的 cloudprovider，它将给当前节点添加一个 Taint 将其标记为不可用。在 cloud-controller-manager 的一个 controller 初始化这个节点后，Kubelet 将删除这个 Taint。

使用这个 alpha 功能特性，结合 tolerationSeconds，Pod 就可以指定当节点出现一个或全部上述问题时，Pod 还能在这个节点上运行多长时间。

比如，一个使用了很多本地状态的应用程序在网络断开时，仍然希望停留在当前节点上运行一段时间，愿意等待网络恢复以避免被驱逐。在这种情况下，Pod 的 Toleration 可以这样配置：

```
tolerations:
- key: "node.alpha.kubernetes.io/unreachable"
  operator: "Exists"
  effect: "NoExecute"
  tolerationSeconds: 6000
```

> **注　意**
>
> Kubernetes 会自动给 Pod 添加一个 key 为 node.kubernetes.io/not-ready 的 Toleration 并配置 tolerationSeconds=300，同样也会给 Pod 添加一个 key 为 node.kubernetes.io/unreachable 的 Toleration 并配置 tolerationSeconds=300，除非用户自定义了上述 key，否则会采用这个默认设置。

这种自动添加 Toleration 的机制保证了在其中一种问题被检测到时，Pod 默认能够继续停留在当前节点运行 5 分钟。这两个默认 Toleration 是由 DefaultTolerationSeconds admission controller 添加的。

DaemonSet 中的 Pod 被创建时，针对以下 Taint 自动添加的 NoExecute 的 Toleration 将不会指

定 tolerationSeconds：
- node.alpha.kubernetes.io/unreachable
- node.kubernetes.io/not-ready

这保证了出现上述问题时 DaemonSet 中的 Pod 永远不会被驱逐。

2.2.16 RBAC

1. RBAC 基本概念

RBAC（Role-Based Access Control，基于角色的访问控制）是一种基于企业内个人用户的角色来管理对计算机或网络资源的访问方法，其在 Kubernetes 1.5 版本中引入，在 1.6 时升级为 Beta 版本，并成为 Kubeadm 安装方式下的默认选项。启用 RBAC 需要在启动 APIServer 时指定 --authorization-mode=RBAC。

RBAC 使用 rbac.authorization.K8S.io API 组来推动授权决策，允许管理员通过 Kubernetes API 动态配置策略。

RBAC API 声明了 4 种顶级资源对象，即 Role、ClusterRole、RoleBinding、ClusterRoleBinding，管理员可以像使用其他 API 资源一样使用 kubectl API 调用这些资源对象。例如：kubectl create -f (resource).yml。

2. Role 和 ClusterRole

Role 和 ClusterRole 的关键区别是，Role 是作用于命名空间内的角色，ClusterRole 作用于整个集群的角色。

在 RBAC API 中，Role 包含表示一组权限的规则。权限纯粹是附加允许的，没有拒绝规则。Role 只能授权对单个命名空间内的资源的访问权限，比如授权对 default 命名空间的读取权限：

```
kind: Role
apiVersion: rbac.authorization.K8S.io/v1
metadata:
  namespace: default
  name: pod-reader
rules:
- apiGroups: [""] # "" indicates the core API group
  resources: ["pods"]
  verbs: ["get", "watch", "list"]
```

ClusterRole 也可将上述权限授予作用于整个集群的 Role，主要区别是，ClusterRole 是集群范围的，因此它们还可以授予对以下内容的访问权限：

- 集群范围的资源（如 Node）。
- 非资源端点（如/healthz）。
- 跨所有命名空间的命名空间资源（如 Pod）。

比如，授予对任何特定命名空间或所有命名空间中的 secret 的读权限（取决于它的绑定方式）：

```
kind: ClusterRole
```

```yaml
apiVersion: rbac.authorization.K8S.io/v1
metadata:
  # "namespace" omitted since ClusterRoles are not namespaced
  name: secret-reader
rules:
- apiGroups: [""]
  resources: ["secrets"]
  verbs: ["get", "watch", "list"]
```

3. RoleBinding 和 ClusterRoleBinding

RoleBinding 将 Role 中定义的权限授予 User、Group 或 Service Account。RoleBinding 和 ClusterRoleBinding 最大的区别与 Role 和 ClusterRole 的区别类似，即 RoleBinding 作用于命名空间，ClusterRoleBinding 作用于集群。

RoleBinding 可以引用同一命名空间的 Role 进行授权，比如将上述创建的 pod-reader 的 Role 授予 default 命名空间的用户 jane，这将允许 jane 读取 default 命名空间中的 Pod：

```yaml
# This role binding allows "jane" to read pods in the "default" namespace.
kind: RoleBinding
apiVersion: rbac.authorization.K8S.io/v1
metadata:
  name: read-pods
  namespace: default
subjects:
- kind: User
  name: jane # Name is case sensitive
  apiGroup: rbac.authorization.K8S.io
roleRef:
  kind: Role #this must be Role or ClusterRole
  name: pod-reader # this must match the name of the Role or ClusterRole you wish to bind to
  apiGroup: rbac.authorization.K8S.io
```

说明

- roleRef：绑定的类别，可以是 Role 或 ClusterRole。

RoleBinding 也可以引用 ClusterRole 来授予对命名空间资源的某些权限。管理员可以为整个集群定义一组公用的 ClusterRole，然后在多个命名空间中重复使用。

比如，创建一个 RoleBinding 引用 ClusterRole，授予 dave 用户读取 development 命名空间的 Secret：

```yaml
# This role binding allows "dave" to read secrets in the "development" namespace.
kind: RoleBinding
apiVersion: rbac.authorization.K8S.io/v1
metadata:
  name: read-secrets
  namespace: development # This only grants permissions within the "development" namespace.
subjects:
- kind: User
  name: dave # Name is case sensitive
  apiGroup: rbac.authorization.K8S.io
```

```
roleRef:
  kind: ClusterRole
  name: secret-reader
  apiGroup: rbac.authorization.K8S.io
```

ClusterRoleBinding 可用于在集群级别和所有命名空间中授予权限，比如允许组 manager 中的所有用户都能读取任何命名空间的 Secret：

```
# This cluster role binding allows anyone in the "manager" group to read secrets
in any namespace.
  kind: ClusterRoleBinding
  apiVersion: rbac.authorization.K8S.io/v1
  metadata:
    name: read-secrets-global
  subjects:
  - kind: Group
    name: manager # Name is case sensitive
    apiGroup: rbac.authorization.K8S.io
  roleRef:
    kind: ClusterRole
    name: secret-reader
    apiGroup: rbac.authorization.K8S.io
```

4. 对集群资源的权限控制

在 Kubernetes 中，大多数资源都由其名称的字符串表示，例如 pods。但是一些 Kubernetes API 涉及的子资源（下级资源），例如 Pod 的日志，对应的 Endpoint 的 URL 是：

```
GET /api/v1/namespaces/{namespace}/pods/{name}/log
```

在这种情况下，pods 是命名空间资源，log 是 Pod 的下级资源，如果对其进行访问控制，要使用斜杠来分隔资源和子资源，比如定义一个 Role 允许读取 Pod 和 Pod 日志：

```
kind: Role
apiVersion: rbac.authorization.K8S.io/v1
metadata:
  namespace: default
  name: pod-and-pod-logs-reader
rules:
- apiGroups: [""]
  resources: ["pods", "pods/log"]
  verbs: ["get", "list"]
```

针对具体资源（使用 resourceNames 指定单个具体资源）的某些请求，也可以通过使用 get、delete、update、patch 等进行授权，比如，只能对一个叫 my-configmap 的 configmap 进行 get 和 update 操作：

```
kind: Role
apiVersion: rbac.authorization.K8S.io/v1
metadata:
  namespace: default
  name: configmap-updater
rules:
- apiGroups: [""]
  resources: ["configmaps"]
```

```
  resourceNames: ["my-configmap"]
  verbs: ["update", "get"]
```

> **注　意**
>
> 如果使用了 resourceNames，则 verbs 不能是 list、watch、create、deletecollection 等。

5. 聚合 ClusterRole

从 Kubernetes 1.9 版本开始，Kubernetes 可以通过一组 ClusterRole 创建聚合 ClusterRoles，聚合 ClusterRoles 的权限由控制器管理，并通过匹配 ClusterRole 的标签自动填充相对应的权限。

比如，匹配 rbac.example.com/aggregate-to-monitoring: "true"标签来创建聚合 ClusterRole：

```
kind: ClusterRole
apiVersion: rbac.authorization.K8S.io/v1
metadata:
  name: monitoring
aggregationRule:
  clusterRoleSelectors:
  - matchLabels:
      rbac.example.com/aggregate-to-monitoring: "true"
rules: [] # Rules are automatically filled in by the controller manager.
```

然后创建与标签选择器匹配的 ClusterRole 向聚合 ClusterRole 添加规则：

```
kind: ClusterRole
apiVersion: rbac.authorization.K8S.io/v1
metadata:
  name: monitoring-endpoints
  labels:
    rbac.example.com/aggregate-to-monitoring: "true"
# These rules will be added to the "monitoring" role.
rules:
- apiGroups: [""]
  resources: ["services", "endpoints", "pods"]
  verbs: ["get", "list", "watch"]
```

6. Role 示例

以下示例允许读取核心 API 组中的资源 Pods（只写了规则 rules 部分）：

```
rules:
- apiGroups: [""]
  resources: ["pods"]
  verbs: ["get", "list", "watch"]
```

允许在 extensions 和 apps API 组中读写 deployments：

```
rules:
- apiGroups: ["extensions", "apps"]
  resources: ["deployments"]
  verbs: ["get", "list", "watch", "create", "update", "patch", "delete"]
```

允许对 Pods 的读和 Job 的读写：

```
rules:
- apiGroups: [""]
```

```
    resources: ["pods"]
    verbs: ["get", "list", "watch"]
  - apiGroups: ["batch", "extensions"]
    resources: ["jobs"]
    verbs: ["get", "list", "watch", "create", "update", "patch", "delete"]
```

允许读取一个名为 my-config 的 ConfigMap（必须绑定到一个 RoleBinding 来限制到一个命名空间下的 ConfigMap）：

```
rules:
- apiGroups: [""]
  resources: ["configmaps"]
  resourceNames: ["my-config"]
  verbs: ["get"]
```

允许读取核心组 Node 资源（Node 属于集群级别的资源，必须放在 ClusterRole 中，并使用 ClusterRoleBinding 进行绑定）：

```
rules:
- apiGroups: [""]
  resources: ["nodes"]
  verbs: ["get", "list", "watch"]
```

允许对非资源端点/healthz 和所有其子资源路径的 Get 和 Post 请求（必须放在 ClusterRole 并与 ClusterRoleBinding 进行绑定）：

```
rules:
- nonResourceURLs: ["/healthz", "/healthz/*"] # '*' in a nonResourceURL is a suffix glob match
  verbs: ["get", "post"]
```

7. RoleBinding 示例

以下示例绑定为名为"alice@example.com"的用户（只显示 subjects 部分）：

```
subjects:
- kind: User
  name: "alice@example.com"
  apiGroup: rbac.authorization.K8S.io
```

绑定为名为"frontend-admins"的组：

```
subjects:
- kind: Group
  name: "frontend-admins"
  apiGroup: rbac.authorization.K8S.io
```

绑定为 kube-system 命名空间中的默认 Service Account：

```
subjects:
- kind: ServiceAccount
  name: default
  namespace: kube-system
```

绑定为 qa 命名空间中的所有 Service Account：

```
subjects:
- kind: Group
```

```
  name: system:serviceaccounts:qa
  apiGroup: rbac.authorization.K8S.io
```

绑定所有 Service Account：

```
subjects:
- kind: Group
  name: system:serviceaccounts
  apiGroup: rbac.authorization.K8S.io
```

绑定所有经过身份验证的用户（v1.5+）：

```
subjects:
- kind: Group
  name: system:authenticated
  apiGroup: rbac.authorization.K8S.io
```

绑定所有未经过身份验证的用户（v1.5+）：

```
subjects:
- kind: Group
  name: system:unauthenticated
  apiGroup: rbac.authorization.K8S.io
```

对于所有用户：

```
subjects:
- kind: Group
  name: system:authenticated
  apiGroup: rbac.authorization.K8S.io
- kind: Group
  name: system:unauthenticated
  apiGroup: rbac.authorization.K8S.io
```

8. 命令行的使用

权限的创建可以使用命令行直接创建，较上述方式更加简单、快捷，下面我们逐一介绍常用命令的使用。

（1）kubectl create role

创建一个 Role，命名为 pod-reader，允许用户在 Pod 上执行 get、watch 和 list：

```
kubectl create role pod-reader --verb=get --verb=list --verb=watch --resource=pods
```

创建一个指定了 resourceNames 的 Role，命名为 pod-reader：

```
kubectl create role pod-reader --verb=get --resource=pods --resource-name=readablepod --resource-name=anotherpod
```

创建一个命名为 foo，并指定 APIGroups 的 Role：

```
kubectl create role foo --verb=get,list,watch --resource=replicasets.apps
```

针对子资源创建一个名为 foo 的 Role：

```
kubectl create role foo --verb=get,list,watch --resource=pods,pods/status
```

针对特定/具体资源创建一个名为 my-component-lease-holder 的 Role：

```
kubectl create role my-component-lease-holder --verb=get,list,watch,update --resource=lease --resource-name=my-component
```

（2）kubectl create clusterrole

创建一个名为 pod-reader 的 ClusterRole，允许用户在 Pod 上执行 get、watch 和 list：

```
kubectl create clusterrole pod-reader --verb=get,list,watch --resource=pods
```

创建一个名为 pod-reader 的 ClusterRole，并指定 resourceName：

```
kubectl create clusterrole pod-reader --verb=get --resource=pods --resource-name=readablepod --resource-name=anotherpod
```

使用指定的 apiGroup 创建一个名为 foo 的 ClusterRole：

```
kubectl create clusterrole foo --verb=get,list,watch --resource=replicasets.apps
```

使用子资源创建一个名为 foo 的 ClusterRole：

```
kubectl create clusterrole foo --verb=get,list,watch --resource=pods,pods/status
```

使用 non-ResourceURL 创建一个名为 foo 的 ClusterRole：

```
kubectl create clusterrole "foo" --verb=get --non-resource-url=/logs/*
```

使用指定标签创建名为 monitoring 的聚合 ClusterRole：

```
kubectl create clusterrole monitoring --aggregation-rule="rbac.example.com/aggregate-to-monitoring=true"
```

（3）kubectl create rolebinding

创建一个名为 bob-admin-binding 的 RoleBinding，将名为 admin 的 ClusterRole 绑定到名为 acme 的命名空间中一个名为 bob 的 user：

```
kubectl create rolebinding bob-admin-binding --clusterrole=admin --user=bob --namespace=acme
```

创建一个名为 myapp-view-binding 的 RoleBinding，将名为 view 的 ClusterRole，绑定到 acme 命名空间中名为 myapp 的 ServiceAccount：

```
kubectl create rolebinding myapp-view-binding --clusterrole=view --serviceaccount=acme:myapp --namespace=acme
```

（4）kubectl create clusterrolebinding

创建一个名为 root-cluster-admin-binding 的 clusterrolebinding，将名为 cluster-admin 的 ClusterRole 绑定到名为 root 的 user：

```
kubectl create clusterrolebinding root-cluster-admin-binding --clusterrole=cluster-admin --user=root
```

创建一个名为 myapp-view-binding 的 clusterrolebinding，将名为 view 的 ClusterRole 绑定到 acme 命名空间中名为 myapp 的 ServiceAccount：

```
kubectl create clusterrolebinding myapp-view-binding --clusterrole=view
--serviceaccount=acme:myapp
```

2.2.17 CronJob

CronJob 用于以时间为基准周期性地执行任务，这些自动化任务和运行在 Linux 或 UNIX 系统上的 CronJob 一样。CronJob 对于创建定期和重复任务非常有用，例如执行备份任务、周期性调度程序接口、发送电子邮件等。

对于 Kubernetes 1.8 以前的版本，需要添加 --runtime-config=batch/v2alpha1=true 参数至 APIServer 中，然后重启 APIServer 和 Controller Manager 用于启用 API，对于 1.8 以后的版本无须修改任何参数，可以直接使用，本节的示例基于 1.8 以上的版本。

1. 创建 CronJob

创建 CronJob 有两种方式，一种是直接使用 kubectl 创建，一种是使用 yaml 文件创建。

使用 kubectl 创建 CronJob 的命令如下：

```
kubectl run hello --schedule="*/1 * * * *" --restart=OnFailure --image=busybox
-- /bin/sh -c "date; echo Hello from the Kubernetes cluster"
```

对应的 yaml 文件如下：

```
apiVersion: batch/v1beta1
kind: CronJob
metadata:
  name: hello
spec:
  schedule: "*/1 * * * *"
  jobTemplate:
    spec:
      template:
        spec:
          containers:
          - name: hello
            image: busybox
            args:
            - /bin/sh
            - -c
            - date; echo Hello from the Kubernetes cluster
          restartPolicy: OnFailure
```

> **说　明**
>
> 本例创建一个每分钟执行一次、打印当前时间和 Hello from the Kubernetes cluster 的计划任务。

查看创建的 CronJob：

```
$ kubectl get cj
NAME    SCHEDULE       SUSPEND   ACTIVE   LAST SCHEDULE   AGE
hello   */1 * * * *    False     0        <none>          5s
```

等待 1 分钟可以查看执行的任务（Jobs）：

```
$ kubectl get jobs
NAME                COMPLETIONS   DURATION   AGE
hello-1558779360    1/1           23s        32s
```

CronJob 每次调用任务的时候会创建一个 Pod 执行命令，执行完任务后，Pod 状态就会变成 Completed，如下所示：

```
$ kubectl get po
NAME                        READY   STATUS      RESTARTS   AGE
hello-1558779360-jcp4r      0/1     Completed   0          37s
```

可以通过 logs 查看 Pod 的执行日志：

```
$ kubectl logs -f hello-1558779360-jcp4r
Sat May 25 10:16:23 UTC 2019
Hello from the Kubernetes cluster
```

如果要删除 CronJob，直接使用 delete 即可：

```
kubectl delete cronjob hello
```

2. 可用参数的配置

定义一个 CronJob 的 yaml 文件如下：

```yaml
apiVersion: v1
items:
- apiVersion: batch/v1beta1
  kind: CronJob
  metadata:
    labels:
      run: hello
    name: hello
    namespace: default
  spec:
    concurrencyPolicy: Allow
    failedJobsHistoryLimit: 1
    jobTemplate:
      metadata:
        creationTimestamp: null
      spec:
        template:
          metadata:
            creationTimestamp: null
            labels:
              run: hello
          spec:
            containers:
            - args:
              - /bin/sh
              - -c
              - date; echo Hello from the Kubernetes cluster
              image: busybox
              imagePullPolicy: Always
              name: hello
```

```yaml
        resources: {}
        terminationMessagePath: /dev/termination-log
        terminationMessagePolicy: File
      dnsPolicy: ClusterFirst
      restartPolicy: OnFailure
      schedulerName: default-scheduler
      securityContext: {}
      terminationGracePeriodSeconds: 30
  schedule: '*/1 * * * *'
  successfulJobsHistoryLimit: 3
  suspend: false
```

其中各参数的说明如下（可以按需修改）：

- schedule　调度周期，和 Linux 一致，分别是分时日月周。
- restartPolicy　重启策略，和 Pod 一致。
- concurrencyPolicy　并发调度策略。可选参数如下：
  - Allow　允许同时运行多个任务。
  - Forbid　不允许并发运行，如果之前的任务尚未完成，新的任务不会被创建。
  - Replace　如果之前的任务尚未完成，新的任务会替换的之前的任务。
- Suspend　如果设置为 true，则暂停后续的任务，默认为 false。
- successfulJobsHistoryLimit　保留多少已完成的任务，按需配置。
- failedJobsHistoryLimit　保留多少失败的任务。

相对于 Linux 上的计划任务，Kubernetes 的 CronJob 更具有可配置性，并且对于执行计划任务的环境只需启动相对应的镜像即可。比如，如果需要 Go 或者 PHP 环境执行任务，就只需要更改任务的镜像为 Go 或者 PHP 即可，而对于 Linux 上的计划任务，则需要安装相对应的执行环境。此外，Kubernetes 的 CronJob 是创建 Pod 来执行，更加清晰明了，查看日志也比较方便。可见，Kubernetes 的 CronJob 更加方便和简单。

更多 CronJob 的内容，可以参考 Kubernetes 的官方文档：https://kubernetes.io/docs/home/。

2.3　小　结

本章讲解了 Docker 和 Kubernetes 在生产环境中常用的基础知识，同时也举例说明了使用场景，希望读者务必深入理解本章内容，因为概念及原理在使用过程中尤为重要，对业务架构的设计和集群排错也都有很大的帮助。有关 Docker 和 Kubernetes 的更多概念可以参考官方文档：

https://kubernetes.io/docs/concepts/

https://docs.docker.com/

第 3 章

Kubernetes 常见应用安装

本章主要介绍容器化部署公司一些常用的应用,在实际使用中,本章的应用采用容器化部署并非必须的,也可自行选择传统的部署方式。

本章的 Ingress 均使用 traefik(可以更改为 nginx-ingress)进行部署,建议首先将之前部署的 Ingress 删除,然后再部署 traefik:

```
openssl req -x509 -nodes -days 365 -newkey rsa:2048 -keyout tls.key -out tls.crt -subj "/CN=K8S-master-lb"
kubectl -n kube-system create secret generic traefik-cert --from-file=tls.key --from-file=tls.crt
kubectl apply -f traefik/
```

3.1 安装 GFS 到 K8S 集群中

在集群中,我们可以使用 GFS、CEPH、NFS 等为 Pod 提供动态持久化存储。本节动态存储主要介绍 GFS 的使用,静态存储方式采用 NFS,其他类型的存储配置类似。

3.1.1 准备工作

为了保证 Pod 能够正常使用 GFS 作为后端存储,需要每台运行 Pod 的节点上提前安装 GFS 的客户端工具,其他存储方式也类似。

所有节点安装 GFS 客户端:

```
yum install glusterfs glusterfs-fuse -y
```

给需要作为 GFS 节点提供存储的节点打上标签:

```
[root@K8S-master01 ~]# kubectl label node K8S-node01 storagenode=glusterfs
node/K8S-node01 labeled
```

```
[root@K8S-master01 ~]# kubectl label node K8S-node02 storagenode=glusterfs
node/K8S-node02 labeled
[root@K8S-master01 ~]# kubectl label node K8S-master01 storagenode=glusterfs
node/K8S-master01 labeled
```

所有节点加载对应模块：

```
[root@K8S-master01 ~]# modprobe dm_snapshot
[root@K8S-master01 ~]# modprobe dm_mirror
[root@K8S-master01 ~]# modprobe dm_thin_pool
```

3.1.2 创建 GFS 集群

这里采用容器化方式部署 GFS 集群，同样也可以使用传统模式部署，在生产环境中，GFS 集群最好是独立于集群之外进行部署，之后只需创建对应的 EndPoints 即可。

本例部署采用 DaemonSet 方式，同时保证已经打上标签的节点上都运行一个 GFS 服务，并且均有提供存储的磁盘。

下载相关安装文件：

```
wget https://github.com/heketi/heketi/releases/download/v7.0.0/
heketi-client-v7.0.0.linux.amd64.tar.gz
```

创建集群：

```
[root@K8S-master01 kubernetes]# pwd
/root/heketi-client/share/heketi/kubernetes
[root@K8S-master01 kubernetes]# kubectl create -f glusterfs-daemonset.json
daemonset.extensions/glusterfs created
[root@K8S-master01 kubernetes]# pwd
/root/heketi-client/share/heketi/kubernetes
```

注意 1：此处采用的是默认的挂载方式，可使用其他磁盘作为 GFS 的工作目录。
注意 2：此处创建的 Namespace 为 default，可按需更改。
注意 3：可使用 gluster/gluster-centos:gluster3u12_centos7 镜像。

查看 GFS Pods：

```
[root@K8S-master01 kubernetes]# kubectl get pods -l glusterfs-node=daemonset
NAME                READY    STATUS     RESTARTS    AGE
glusterfs-5npwn     1/1      Running    0           1m
glusterfs-bd5dx     1/1      Running    0           1m
...
```

3.1.3 创建 Heketi 服务

Heketi 是一个提供 RESTful API 管理 GFS 卷的框架，能够在 Kubernetes、OpenShift、OpenStack 等云平台上实现动态存储资源供应，支持 GFS 多集群管理，便于管理员对 GFS 进行操作，在 Kubernetes 集群中，Pod 将存储的请求发送至 Heketi，然后 Heketi 控制 GFS 集群创建对应的存储卷。

创建 Heketi 的 ServiceAccount 对象：

```
[root@K8S-master01 kubernetes]# cat heketi-service-account.json
{
  "apiVersion": "v1",
  "kind": "ServiceAccount",
  "metadata": {
    "name": "heketi-service-account"
  }
}
[root@K8S-master01 kubernetes]# kubectl create -f heketi-service-account.json
serviceaccount/heketi-service-account created
[root@K8S-master01 kubernetes]# pwd
/root/heketi-client/share/heketi/kubernetes
[root@K8S-master01 kubernetes]# kubectl get sa
NAME                       SECRETS   AGE
default                    1         13d
heketi-service-account     1         <invalid>
```

创建 Heketi 对应的权限和 Secret：

```
[root@K8S-master01 kubernetes]# kubectl create clusterrolebinding
heketi-gluster-admin --clusterrole=edit
--serviceaccount=default:heketi-service-account
    clusterrolebinding.rbac.authorization.K8S.io/heketi-gluster-admin created
    [root@K8S-master01 kubernetes]# kubectl create secret generic
heketi-config-secret --from-file=./heketi.json
    secret/heketi-config-secret created
```

初始化部署 Heketi：

```
[root@K8S-master01 kubernetes]# kubectl create -f heketi-bootstrap.json
secret/heketi-db-backup created
service/heketi created
deployment.extensions/heketi created
[root@K8S-master01 kubernetes]# pwd
/root/heketi-client/share/heketi/kubernetes
```

3.1.4 创建 GFS 集群

本节使用 Heketi 创建 GFS 集群，其管理方式更加简单和高效。

复制 heketi-cli 至 /usr/local/bin：

```
[root@K8S-master01 heketi-client]# cp bin/heketi-cli /usr/local/bin/
[root@K8S-master01 heketi-client]# pwd
/root/heketi-client

[root@K8S-master01 heketi-client]# heketi-cli -v
heketi-cli v7.0.0
```

修改 topology-sample，manage 为 GFS 管理服务的节点（Node）主机名，storage 为节点的 IP 地址，devices 为节点上的裸设备，也就是用于提供存储的磁盘最好使用裸设备：

```
[root@K8S-master01 kubernetes]# cat topology-sample.json
```

```json
{
    "clusters": [
      {
        "nodes": [
          {
            "node": {
              "hostnames": {
                "manage": [
                  "K8S-master01"
                ],
                "storage": [
                  "192.168.20.20"
                ]
              },
              "zone": 1
            },
            "devices": [
              {
                "name": "/dev/sdc",
                "destroydata": false
              }
            ]
          },
          {
            "node": {
              "hostnames": {
                "manage": [
                  "K8S-node01"
                ],
                "storage": [
                  "192.168.20.30"
                ]
              },
              "zone": 1
            },
            "devices": [
              {
                "name": "/dev/sdb",
                "destroydata": false
              }
            ]
          },
          {
            "node": {
              "hostnames": {
                "manage": [
                  "K8S-node02"
                ],
                "storage": [
                  "192.168.20.31"
                ]
              },
              "zone": 1
            },
            "devices": [
```

```
        {
          "name": "/dev/sdb",
          "destroydata": false
        }
      ]
    }
  ]
}
```

查看当前 Heketi 的 ClusterIP：

```
[root@K8S-master01 kubernetes]# kubectl get svc | grep heketi
deploy-heketi    ClusterIP    10.110.217.153    <none>    8080/TCP    26m
[root@K8S-master01kubernetes]#export
HEKETI_CLI_SERVER=http://10.110.217.153:8080
```

使用 Heketi 创建 GFS 集群：

```
[root@K8S-master01 kubernetes]# heketi-cli topology load
--json=topology-sample.json
    Creating cluster ... ID: a058723afae149618337299c84a1eaed
        Allowing file volumes on cluster.
        Allowing block volumes on cluster.
        Creating node K8S-master01 ... ID: 929909065ceedb59c1b9c235fc3298ec
            Adding device /dev/sdc ... OK
        Creating node K8S-node01 ... ID: 37409d82b9ef27f73ccc847853eec429
            Adding device /dev/sdb ... OK
        Creating node K8S-node02 ... ID: e3ab676be27945749bba90efb34f2eb9
            Adding device /dev/sdb ... OK
```

之前创建的 Heketi 未配置持久化卷，如果 Heketi 的 Pod 重启，可能会丢失之前的配置信息，所以现在创建 Heketi 持久化卷，对 Heketi 的数据进行持久化。该持久化方式采用 GFS 提供的动态存储，也可以采用其他方式进行持久化：

```
#所有节点安装device-mapper*
[root@K8S-master01 kubernetes]# yum install device-mapper* -y
[root@K8S-master01 kubernetes]# heketi-cli setup-openshift-heketi-storage
Saving heketi-storage.json
[root@K8S-master01 kubernetes]# ls
glusterfs-daemonset.json   heketi.json                  heketi-storage.json
heketi-bootstrap.json      heketi-service-account.json  README.md
heketi-deployment.json     heketi-start.sh              topology-sample.json
[root@K8S-master01 kubernetes]# kubectl create -f heketi-storage.json
secret/heketi-storage-secret created
endpoints/heketi-storage-endpoints created
service/heketi-storage-endpoints created
job.batch/heketi-storage-copy-job created
```

如果出现如下报错信息，就在所有节点执行 modprobe dm_thin_pool，重新执行此命令即可：

```
[root@K8S-master01 kubernetes]# heketi-cli setup-openshift-heketi-storage
Error: /usr/sbin/modprobe failed: 1
  thin: Required device-mapper target(s) not detected in your kernel.
  Run `lvcreate --help' for more information.
```

删除中间产物：

```
[root@K8S-master01 kubernetes]# kubectl delete all,service,jobs,deployment,secret --selector="deploy-heketi"
pod "deploy-heketi-59f8dbc97f-5rf6s" deleted
service "deploy-heketi" deleted
service "heketi" deleted
deployment.apps "deploy-heketi" deleted
replicaset.apps "deploy-heketi-59f8dbc97f" deleted
job.batch "heketi-storage-copy-job" deleted
secret "heketi-storage-secret" deleted
```

创建持久化 Heketi，持久化方式也可以选择其他方式：

```
[root@K8S-master01 kubernetes]# kubectl create -f heketi-deployment.json
service/heketi created
deployment.extensions/heketi created
```

待 Pod 全部启动，即表示创建成功：

```
[root@K8S-master01 kubernetes]# kubectl get po
NAME                          READY   STATUS    RESTARTS   AGE
glusterfs-5npwn               1/1     Running   0          3h
glusterfs-8zfzq               1/1     Running   0          3h
glusterfs-bd5dx               1/1     Running   0          3h
heketi-5cb5f55d9f-5mtqt       1/1     Running   0          2m
```

查看最新部署的持久化 Heketi 的 svc，并更改 HEKETI_CLI_SERVER 的值：

```
[root@K8S-master01 kubernetes]# kubectl get svc
NAME                        TYPE        CLUSTER-IP      EXTERNAL-IP   PORT(S)    AGE
heketi                      ClusterIP   10.111.95.240   <none>        8080/TCP   12h
heketi-storage-endpoints    ClusterIP   10.99.28.153    <none>        1/TCP      12h
kubernetes                  ClusterIP   10.96.0.1       <none>        443/TCP    14d
[root@K8S-master01 kubernetes]# export HEKETI_CLI_SERVER=http://10.111.95.240:8080
[root@K8S-master01 kubernetes]# curl http://10.111.95.240:8080/hello
Hello from Heketi
```

查看 GFS 集群信息，其他操作可参考 Heketi 官方文档：

```
[root@K8S-master01 kubernetes]# heketi-cli topology info

Cluster Id: 5dec5676c731498c2bdf996e110a3e5e

    File:  true
    Block: true

    Volumes:

    Name: heketidbstorage
    Size: 2
    Id: 828dc2dfaa00b7213e831b91c6213ae4
```

```
        Cluster Id: 5dec5676c731498c2bdf996e110a3e5e
        Mount: 192.168.20.31:heketidbstorage
        Mount Options: backup-volfile-servers=192.168.20.30,192.168.20.20
        Durability Type: replicate
        Replica: 3
        Snapshot: Disabled

        Bricks:
            Id: 16b7270d7db1b3cfe9656b64c2a3916c
            Path: /var/lib/heketi/mounts/vg_04290ec786dc7752a469b66f5e94458f/
brick_16b7270d7db1b3cfe9656b64c2a3916c/brick
            Size (GiB): 2
            Node: fb181b0cef571e9af7d84d2ecf534585
            Device: 04290ec786dc7752a469b66f5e94458f

            Id: 828da093d9d78a2b1c382b13cc4da4a1
            Path: /var/lib/heketi/mounts/vg_80b61df999fcac26ebca6e28c4da8e61/
brick_828da093d9d78a2b1c382b13cc4da4a1/brick
            Size (GiB): 2
            Node: d38819746cab7d567ba5f5f4fea45d91
            Device: 80b61df999fcac26ebca6e28c4da8e61

            Id: e8ef0e68ccc3a0416f73bc111cffee61
            Path: /var/lib/heketi/mounts/vg_82af8e5f2fb2e1396f7c9e9f7698a178/
brick_e8ef0e68ccc3a0416f73bc111cffee61/brick
            Size (GiB): 2
            Node: 0f00835397868d3591f45432e432ba38
            Device: 82af8e5f2fb2e1396f7c9e9f7698a178


    Nodes:

    Node Id: 0f00835397868d3591f45432e432ba38
    State: online
    Cluster Id: 5dec5676c731498c2bdf996e110a3e5e
    Zone: 1
    Management Hostnames: K8S-node02
    Storage Hostnames: 192.168.20.31
    Devices:
        Id:82af8e5f2fb2e1396f7c9e9f7698a178   Name:/dev/sdb
State:online    Size (GiB):39     Used (GiB):22      Free (GiB):17
            Bricks:
                Id:e8ef0e68ccc3a0416f73bc111cffee61   Size (GiB):2    Path:
/var/lib/heketi/mounts/vg_82af8e5f2fb2e1396f7c9e9f7698a178/brick_e8ef0e68ccc3a
0416f73bc111cffee61/brick

    Node Id: d38819746cab7d567ba5f5f4fea45d91
    State: online
    Cluster Id: 5dec5676c731498c2bdf996e110a3e5e
    Zone: 1
    Management Hostnames: K8S-node01
    Storage Hostnames: 192.168.20.30
    Devices:
        Id:80b61df999fcac26ebca6e28c4da8e61  Name:/dev/sdb
State:online    Size (GiB):39     Used (GiB):22      Free (GiB):17
```

```
            Bricks:
                Id:828da093d9d78a2b1c382b13cc4da4a1    Size (GiB):2      Path:
/var/lib/heketi/mounts/vg_80b61df999fcac26ebca6e28c4da8e61/brick_828da093d9d78
a2b1c382b13cc4da4a1/brick

        Node Id: fb181b0cef571e9af7d84d2ecf534585
        State: online
        Cluster Id: 5dec5676c731498c2bdf996e110a3e5e
        Zone: 1
        Management Hostnames: K8S-master01
        Storage Hostnames: 192.168.20.20
        Devices:
            Id:04290ec786dc7752a469b66f5e94458f   Name:/dev/sdc
State:online    Size (GiB):39    Used (GiB):22    Free (GiB):17
            Bricks:
                Id:16b7270d7db1b3cfe9656b64c2a3916c    Size (GiB):2      Path:
/var/lib/heketi/mounts/vg_04290ec786dc7752a469b66f5e94458f/brick_16b7270d7db1b
3cfe9656b64c2a3916c/brick
```

3.1.5 创建 StorageClass

提前创建 StorageClass，然后 Pod 直接在 StorageClass 选项配置选择该 StorageClass，即可通过该 StorageClass 创建对应的 PV。

创建 StorageClass 文件如下：

```
[root@K8S-master01 gfs]# cat storageclass-gfs-heketi.yaml
apiVersion: storage.K8S.io/v1
kind: StorageClass
metadata:
  name: gluster-heketi
provisioner: kubernetes.io/glusterfs
parameters:
  resturl: "http://10.111.95.240:8080"
  restauthenabled: "false"
```

说明

Provisioner 参数须设置为"kubernetes.io/glusterfs"。

resturl 地址为 API Server 所在主机可以访问到的 Heketi 服务的某个地址。

其他详细参数可以参考 2.2.12 节。

3.1.6 测试使用 GFS 动态存储

创建一个 Pod 使用动态 PV，在 storageClassName 指定之前创建的 StorageClass 的名字，即 gluster-heketi：

```
[root@K8S-master01 gfs]# cat pod-use-pvc.yaml
apiVersion: v1
kind: Pod
metadata:
```

```yaml
    name: pod-use-pvc
spec:
  containers:
  - name: pod-use-pvc
    image: busybox
    command:
    - sleep
    - "3600"
    volumeMounts:
    - name: gluster-volume
      mountPath: "/pv-data"
      readOnly: false
  volumes:
  - name: gluster-volume
    persistentVolumeClaim:
      claimName: pvc-gluster-heketi
---
kind: PersistentVolumeClaim
apiVersion: v1
metadata:
  name: pvc-gluster-heketi
spec:
  accessModes: [ "ReadWriteOnce" ]
  storageClassName: "gluster-heketi"
  resources:
    requests:
      storage: 1Gi
```

创建 Pod：

```
[root@K8S-master01 gfs]# kubectl create -f pod-use-pvc.yaml
pod/pod-use-pvc created
```

PVC 定义一旦生成，系统便触发 Heketi 进行相应的操作，主要为在 GFS 集群上创建 brick，再创建并启动一个卷（volume）。

创建的 PV 及 PVC 如下：

```
[root@K8S-master01 gfs]# kubectl get pv,pvc | grep gluster
    persistentvolume/pvc-4a8033e8-e7f7-11e8-9a09-000c293bfe27   1Gi        RWO
Delete           Bound       default/pvc-gluster-heketi
gluster-heketi           5m
    persistentvolumeclaim/pvc-gluster-heketi   Bound
pvc-4a8033e8-e7f7-11e8-9a09-000c293bfe27   1Gi        RWO
gluster-heketi   5m
```

3.1.7 测试数据

测试使用该 PV 的 Pod 之间能否共享数据。

进入到 Pod 并创建文件：

```
[root@K8S-master01 /]# kubectl exec -ti pod-use-pvc -- /bin/sh
/ # cd /pv-data/
```

```
/pv-data # mkdir {1..10}
/pv-data # ls
{1..10}
```

查看创建的卷（volume）：

```
[root@K8S-master01 /]# heketi-cli topology info
Cluster Id: 5dec5676c731498c2bdf996e110a3e5e

    File:  true
    Block: true

    Volumes:

    Name: vol_56d636b452d31a9d4cb523d752ad0891
    Size: 1
    Id: 56d636b452d31a9d4cb523d752ad0891
    Cluster Id: 5dec5676c731498c2bdf996e110a3e5e
    Mount: 192.168.20.31:vol_56d636b452d31a9d4cb523d752ad0891
    Mount Options: backup-volfile-servers=192.168.20.30,192.168.20.20
    Durability Type: replicate
    Replica: 3
    Snapshot: Enabled
...
...
```

或者使用 volume list 查看：

```
[root@K8S-master01 mnt]# heketi-cli volume list
    Id:56d636b452d31a9d4cb523d752ad0891
Cluster:5dec5676c731498c2bdf996e110a3e5e
Name:vol_56d636b452d31a9d4cb523d752ad0891
    Id:828dc2dfaa00b7213e831b91c6213ae4
Cluster:5dec5676c731498c2bdf996e110a3e5e Name:heketidbstorage
    [root@K8S-master01 mnt]#
```

查看数据。其中 vol_56d636b452d31a9d4cb523d752ad0891 为卷名（Volume Name）挂载：

```
[root@K8S-master01 /]# mount -t glusterfs
192.168.20.31:vol_56d636b452d31a9d4cb523d752ad0891  /mnt/
[root@K8S-master01 /]# cd /mnt/
[root@K8S-master01 mnt]# ls
{1..10}
```

3.1.8　测试 Deployment

测试在 Deployment 部署方式下是否能够正常使用 StorageClass。

创建一个 Nginx 的 Deployment 如下：

```
[root@K8S-master01 gfs]# cat nginx-gluster.yaml
apiVersion: extensions/v1beta1
kind: Deployment
metadata:
  name: nginx-gfs
spec:
```

```yaml
      replicas: 2
      template:
        metadata:
          labels:
            name: nginx
        spec:
          containers:
            - name: nginx
              image: nginx
              imagePullPolicy: IfNotPresent
              ports:
                - containerPort: 80
              volumeMounts:
                - name: nginx-gfs-html
                  mountPath: "/usr/share/nginx/html"
                - name: nginx-gfs-conf
                  mountPath: "/etc/nginx/conf.d"
          volumes:
          - name: nginx-gfs-html
            persistentVolumeClaim:
              claimName: glusterfs-nginx-html
          - name: nginx-gfs-conf
            persistentVolumeClaim:
              claimName: glusterfs-nginx-conf
---
kind: PersistentVolumeClaim
apiVersion: v1
metadata:
  name: glusterfs-nginx-html
spec:
  accessModes: [ "ReadWriteMany" ]
  storageClassName: "gluster-heketi"
  resources:
    requests:
      storage: 500Mi
---
kind: PersistentVolumeClaim
apiVersion: v1
metadata:
  name: glusterfs-nginx-conf
spec:
  accessModes: [ "ReadWriteMany" ]
  storageClassName: "gluster-heketi"
  resources:
    requests:
      storage: 10Mi
```

上述例子为了演示使用了两个 PVC，实际环境可以使用 subPath 来区分 conf 和 html，当然也可以直接指定卷，此时不单独创建 PVC。

直接创建的方式如下：

```yaml
spec:
  volumeClaimTemplates:
  - metadata:
      name: rabbitmq-storage
```

```yaml
  spec:
    accessModes:
    - ReadWriteOnce
    storageClassName: "gluster-heketi"
    resources:
      requests:
        storage: 4Gi
```

查看资源：

```
[root@K8S-master01 gfs]# kubectl get po,pvc,pv | grep nginx
pod/nginx-gfs-77c758ccc-2hwl6              1/1       Running             0       4m
pod/nginx-gfs-77c758ccc-kxzfz              0/1       ContainerCreating   0       3m

persistentvolumeclaim/glusterfs-nginx-conf   Bound   pvc-f40c5d4b-e800-11e8-8a89-000c293ad492   1Gi   RWX   gluster-heketi   2m
persistentvolumeclaim/glusterfs-nginx-html   Bound   pvc-f40914f8-e800-11e8-8a89-000c293ad492   1Gi   RWX   gluster-heketi   2m

persistentvolume/pvc-f40914f8-e800-11e8-8a89-000c293ad492   1Gi   RWX   Delete   Bound   default/glusterfs-nginx-html   gluster-heketi   4m
persistentvolume/pvc-f40c5d4b-e800-11e8-8a89-000c293ad492   1Gi   RWX   Delete   Bound   default/glusterfs-nginx-conf   gluster-heketi   4m
```

查看挂载情况：

```
[root@K8S-master01 gfs]# kubectl exec -ti nginx-gfs-77c758ccc-2hwl6 -- df -Th
Filesystem                                        Type            Size  Used Avail Use% Mounted on
overlay                                           overlay          86G  6.6G   80G   8% /
tmpfs                                             tmpfs           7.8G     0  7.8G   0% /dev
tmpfs                                             tmpfs           7.8G     0  7.8G   0% /sys/fs/cgroup
/dev/mapper/centos-root                           xfs              86G  6.6G   80G   8% /etc/hosts
shm                                               tmpfs            64M     0   64M   0% /dev/shm
192.168.20.20:vol_b9c68075c6f20438b46db892d15ed45a fuse.glusterfs 1014M   43M  972M   5% /etc/nginx/conf.d
192.168.20.20:vol_32146a51be9f980c14bc86c34f67ebd5 fuse.glusterfs 1014M   43M  972M   5% /usr/share/nginx/html
tmpfs
```

宿主机挂载并创建 index.html：

```
[root@K8S-master01 gfs]# mount -t glusterfs 192.168.20.20:vol_32146a51be9f980c14bc86c34f67ebd5 /mnt/
[root@K8S-master01 gfs]# cd /mnt/
[root@K8S-master01 mnt]# ls
[root@K8S-master01 mnt]# echo "test" > index.html
[root@K8S-master01 mnt]# kubectl exec -ti nginx-gfs-77c758ccc-2hwl6 -- cat /usr/share/nginx/html/index.html
test
```

扩容 Nginx，查看是否能正常挂载目录：

```
[root@K8S-master01 ~]# kubectl get deploy
NAME        DESIRED   CURRENT   UP-TO-DATE   AVAILABLE   AGE
```

```
heketi           1        1        1        1          14h
nginx-gfs        2        2        2        2          23m
[root@K8S-master01 ~]# kubectl scale deploy nginx-gfs --replicas 3
deployment.extensions/nginx-gfs scaled
[root@K8S-master01 ~]# kubectl get po
NAME                              READY     STATUS      RESTARTS    AGE
glusterfs-5npwn                   1/1       Running     0           18h
glusterfs-8zfzq                   1/1       Running     0           17h
glusterfs-bd5dx                   1/1       Running     0           18h
heketi-5cb5f55d9f-5mtqt           1/1       Running     0           14h
nginx-gfs-77c758ccc-2hwl6         1/1       Running     0           11m
nginx-gfs-77c758ccc-6fphl         1/1       Running     0           8m
nginx-gfs-77c758ccc-kxzfz         1/1       Running     0           10m
```

查看文件内容：

```
[root@K8S-master01 ~]# kubectl exec -ti nginx-gfs-77c758ccc-6fphl -- cat
/usr/share/nginx/html/index.html
 test
```

3.2 安装 Helm 到 K8S 集群中

对于复杂的应用中间件需要设置镜像运行的需求、环境变量，并且需要定制存储、网络等设置，以及设计和编写 Deployment、Configmap、Service 及 Ingress 等相关 yaml 配置文件，再提交给 Kubernetes 进行部署，这些复杂的过程正在逐步被 Helm 应用包管理工具所实现。

3.2.1 基本概念

Helm 是一个由 CNCF 孵化和管理的项目，用于对需要在 K8S 上部署复杂应用进行定义、安装和更新。Helm 以 Chart 的方式对应用软件进行描述，可以方便地创建、版本化、共享和发布复杂的应用软件。

使用 Helm 会涉及以下几个术语：

- Chart　一个 Helm 包，其中包含了运行一个应用所需要的工具和资源定义，还可能包含 Kubernetes 集群中的服务定义，类似于 Homebrew 中的 formula、apt 中的 dpkg 或者 yum 中的 rpm 文件。
- Release　在 K8S 集群上运行一个 Chart 实例。在同一个集群上，一个 Chart 可以安装多次，例如有一个 MySQL Chart，如果想在服务器上运行两个数据库，就可以基于这个 Chart 安装两次。每次安装都会生成新的 Release，并有独立的 Release 名称。
- Repository　用于存放和共享 Chart 的仓库。

简单来说，Helm 的任务是在仓库中查找需要的 Chart，然后将 Chart 以 Release 的形式安装到 K8S 集群中。

3.2.2 安装 Helm

Helm 由以下两个组件组成：

- HelmClinet 客户端，拥有对 Repository、Chart、Release 等对象的管理能力。
- TillerServer 负责客户端指令和 K8S 集群之间的交互，根据 Chart 的定义生成和管理各种 K8S 的资源对象。

可以通过二进制文件或脚本方式安装 HelmClient。

下载最新版二进制文件并解压缩，下载地址为：https://github.com/helm/helm/releases

```
[root@K8S-master01 ~]# tar xf helm-v2.11.0-linux-amd64.tar.gz
[root@K8S-master01 ~]# cp linux-amd64/helm linux-amd64/tiller /usr/local/bin/
```

所有节点下载 tiller:v[helm-version]镜像，helm-version 为上述 Helm 的版本 2.11.0，安装 TillerServer：

```
docker pull dotbalo/tiller:v2.11.0
```

使用 helm init 安装 tiller：

```
[root@K8S-master01 ~]# helm init --tiller-image dotbalo/tiller:v2.11.0
Creating /root/.helm
Creating /root/.helm/repository
Creating /root/.helm/repository/cache
Creating /root/.helm/repository/local
Creating /root/.helm/plugins
Creating /root/.helm/starters
Creating /root/.helm/cache/archive
Creating /root/.helm/repository/repositories.yaml
Adding stable repo with URL: https://kubernetes-charts.storage.googleapis.com
Adding local repo with URL: http://127.0.0.1:8879/charts
$HELM_HOME has been configured at /root/.helm.

Tiller (the Helm server-side component) has been installed into your Kubernetes Cluster.

Please note: by default, Tiller is deployed with an insecure 'allow unauthenticated users' policy.
To prevent this, run `helm init` with the --tiller-tls-verify flag.
For more information on securing your installation see:
https://docs.helm.sh/using_helm/#securing-your-helm-installation
Happy Helming!
```

创建 Helm 的 ServiceAccount：

```
kubectl create serviceaccount --namespace kube-system tiller

kubectl create clusterrolebinding tiller-cluster-rule --clusterrole=cluster-admin --serviceaccount=kube-system:tiller

kubectl patch deploy --namespace kube-system tiller-deploy -p
```

```
'{"spec":{"template":{"spec":{"serviceAccount":"tiller"}}}}'
```

查看 helm version 及 Pod 的状态:

```
[root@K8S-master01 ~]# helm version
Client: &version.Version{SemVer:"v2.11.0",
GitCommit:"2e55dbe1fdb5fdb96b75ff144a339489417b146b", GitTreeState:"clean"}
Server: &version.Version{SemVer:"v2.11.0",
GitCommit:"2e55dbe1fdb5fdb96b75ff144a339489417b146b", GitTreeState:"clean"}
[root@K8S-master01 ~]# kubectl get pod -n kube-system | grep tiller
tiller-deploy-5d7c8fcd59-d4djx           1/1       Running   0          49s
[root@K8S-master01 ~]# kubectl get pod,svc -n kube-system | grep tiller
pod/tiller-deploy-5d7c8fcd59-d4djx       1/1       Running   0          3m

service/tiller-deploy            ClusterIP   10.106.28.190    <none>
44134/TCP        5m
```

3.2.3　Helm 的使用

helm search 用于搜索可用的 Chart, Helm 初始化完成之后, 默认配置为使用官方的 K8S chart 仓库, 通过 search 查找可用的 Chart:

```
[root@K8S-master01 ~]# helm search gitlab
NAME                  CHART VERSION    APP VERSION    DESCRIPTION
stable/gitlab-ce      0.2.2            9.4.1          GitLab Community Edition
stable/gitlab-ee      0.2.2            9.4.1          GitLab Enterprise Edition
[root@K8S-master01 ~]# helm search | more
NAME                         CHART VERSION    APP VERSION    DESCRIPTION

stable/acs-engine-autoscaler   2.2.0     2.1.1       Scales worker n
odes within agent pools
stable/aerospike               0.1.7     v3.14.1.2   A Helm chart fo
r Aerospike in Kubernetes
stable/anchore-engine          0.9.0     0.3.0       Anchore contain
er analysis and policy evaluation engine s...
stable/apm-server              0.1.0     6.2.4       The server rece
ives data from the Elastic APM agents and ...
stable/ark                     1.2.2     0.9.1       A Helm chart fo
r ark
stable/artifactory             7.3.1     6.1.0       DEPRECATED Univ
ersal Repository Manager supporting all ma...
stable/artifactory-ha          0.4.1     6.2.0       DEPRECATED Univ
ersal Repository Manager supporting all ma...
stable/auditbeat               0.3.1     6.4.3       A lightweight s
hipper to audit the activities of users an...
--More--
```

查看详细信息:

```
[root@K8S-master01 ~]# helm search gitlab
NAME                CHART VERSION    APP VERSION    DESCRIPTION
stable/gitlab-ce    0.2.2            9.4.1          GitLab Community Edition
stable/gitlab-ee    0.2.2            9.4.1          GitLab Enterprise Edition
[root@K8S-master01 ~]# helm inspect stable/gitlab-ce
```

下载一个 Chart：

```
helmfetch ChartName
```

3.3　安装 Redis 集群模式到 K8S 集群中

本节演示安装 Redis 4.0.8 集群模式到 K8S 中，本小节安装采用的是 NFS（阿里云可采用 NAS）作为持久化存储，当然也可以使用上节创建的 GFS 提供动态存储，只需更改 redis-cluster-ss.yaml 的 storageClassName 即可，同时无须再创建 redis-cluster-pv.yaml 文件。在生产环境中，对 Redis 集群实现持久化部署并不是必须的，可以采用 hostPath 模式将宿主机的本地目录挂载至 Redis 存储目录，再加上 Pod 互斥，不让 Redis 实例部署在同一个宿主机上，之后再利用节点亲和力，尽量将 Redis 实例部署在原有的宿主机上，此种方式和直接在宿主机上部署 Redis 并无太大区别，并且实现了 Redis 的自动容灾功能。当然，在实际使用时，也可以不对 Redis 进行持久化部署，因为生产环境一般采用 Cluster 模式部署，同时宕机的可能性较小。

3.3.1　各文件介绍

1．redis-cluster-configmap.yaml 文件

使用 ConfigMap 配置 Redis 的配置文件，请按需修改：

```yaml
kind: ConfigMap
apiVersion: v1
metadata:
  name: redis-cluster-config
  namespace: public-service
  labels:
    addonmanager.kubernetes.io/mode: Reconcile
data:
  redis-cluster.conf: |
    # 节点端口
    port 6379
    # # 开启集群模式
    cluster-enabled yes
    # # 节点超时时间，单位毫秒
    cluster-node-timeout 15000
    # # 集群内部配置文件
    cluster-config-file "nodes.conf"
```

2．redis-cluster-pv.yaml 文件

定义 Redis 的持久化文件，请按需修改：

```yaml
apiVersion: v1
kind: PersistentVolume
metadata:
  name: pv-redis-cluster-1
  namespace: public-service
```

```yaml
spec:
  capacity:
    storage: 1Gi
  accessModes:
    - ReadWriteOnce
  volumeMode: Filesystem
  persistentVolumeReclaimPolicy: Recycle
  storageClassName: "redis-cluster-storage-class"
  nfs:
    # real share directory
    path: /K8S/redis-cluster/1
    # nfs real ip
    server: 192.168.2.2

---
apiVersion: v1
kind: PersistentVolume
metadata:
  name: pv-redis-cluster-2
  namespace: public-service
spec:
  capacity:
    storage: 1Gi
  accessModes:
    - ReadWriteOnce
  volumeMode: Filesystem
  persistentVolumeReclaimPolicy: Recycle
  storageClassName: "redis-cluster-storage-class"
  nfs:
    # real share directory
    path: /K8S/redis-cluster/2
    # nfs real ip
    server: 192.168.2.2

---
apiVersion: v1
kind: PersistentVolume
metadata:
  name: pv-redis-cluster-3
  namespace: public-service
spec:
  capacity:
    storage: 1Gi
  accessModes:
    - ReadWriteOnce
  volumeMode: Filesystem
  persistentVolumeReclaimPolicy: Recycle
  storageClassName: "redis-cluster-storage-class"
  nfs:
    # real share directory
    path: /K8S/redis-cluster/3
    # nfs real ip
    server: 192.168.2.2

---
```

```yaml
apiVersion: v1
kind: PersistentVolume
metadata:
  name: pv-redis-cluster-4
  namespace: public-service
spec:
  capacity:
    storage: 1Gi
  accessModes:
    - ReadWriteOnce
  volumeMode: Filesystem
  persistentVolumeReclaimPolicy: Recycle
  storageClassName: "redis-cluster-storage-class"
  nfs:
    # real share directory
    path: /K8S/redis-cluster/4
    # nfs real ip
    server: 192.168.2.2

---
apiVersion: v1
kind: PersistentVolume
metadata:
  name: pv-redis-cluster-5
  namespace: public-service
spec:
  capacity:
    storage: 1Gi
  accessModes:
    - ReadWriteOnce
  volumeMode: Filesystem
  persistentVolumeReclaimPolicy: Recycle
  storageClassName: "redis-cluster-storage-class"
  nfs:
    # real share directory
    path: /K8S/redis-cluster/5
    # nfs real ip
    server: 192.168.2.2

---
apiVersion: v1
kind: PersistentVolume
metadata:
  name: pv-redis-cluster-6
  namespace: public-service
spec:
  capacity:
    storage: 1Gi
  accessModes:
    - ReadWriteOnce
  volumeMode: Filesystem
  persistentVolumeReclaimPolicy: Recycle
  storageClassName: "redis-cluster-storage-class"
  nfs:
    # real share directory
```

```
      path: /K8S/redis-cluster/6
      # nfs real ip
      server: 192.168.2.2
```

此文件为 Redis 集群 6 个实例所用的 PV 文件,动态存储无须创建此文件,其中 path 是 NFS 共享目录,Server 为 NFS 服务器地址,按需修改。

3. redis-cluster-rbac.yaml 文件

此文件定义的是一些权限,无须更改。

```
apiVersion: v1
kind: ServiceAccount
metadata:
  name: redis-cluster
  namespace: public-service
---
kind: Role
apiVersion: rbac.authorization.K8S.io/v1beta1
metadata:
  name: redis-cluster
  namespace: public-service
rules:
  - apiGroups:
      - ""
    resources:
      - endpoints
    verbs:
      - get
---
kind: RoleBinding
apiVersion: rbac.authorization.K8S.io/v1beta1
metadata:
  name: redis-cluster
  namespace: public-service
roleRef:
  apiGroup: rbac.authorization.K8S.io
  kind: Role
  name: redis-cluster
subjects:
- kind: ServiceAccount
  name: redis-cluster
  namespace: public-service
```

4. redis-cluster-service.yaml 文件

此文件用来定义 Redis 的 service,用于集群节点的通信。

```
kind: Service
apiVersion: v1
metadata:
  labels:
    app: redis-cluster-ss
  name: redis-cluster-ss
  namespace: public-service
spec:
```

```yaml
  clusterIP: None
  ports:
  - name: redis
    port: 6379
    targetPort: 6379
  selector:
    app: redis-cluster-ss
```

如果开发或运维需要连接到该集群可以使用 NodePort，程序连接直接使用 SVC 地址即可。

5. redis-cluster-ss.yaml 文件

本例 Redis 的安装采用 StatefulSet 模式，redis-cluster-ss.yaml 文件定义如下：

```yaml
kind: StatefulSet
apiVersion: apps/v1beta1
metadata:
  labels:
    app: redis-cluster-ss
  name: redis-cluster-ss
  namespace: public-service
spec:
  replicas: 6
  selector:
    matchLabels:
      app: redis-cluster-ss
  serviceName: redis-cluster-ss
  template:
    metadata:
      labels:
        app: redis-cluster-ss
    spec:
      containers:
      - args:
        - -c
        - redis-server /mnt/redis-cluster.conf
        command:
        - sh
        image: dotbalo/redis-trib:4.0.10
        imagePullPolicy: IfNotPresent
        name: redis-cluster
        ports:
        - containerPort: 6379
          name: masterport
          protocol: TCP
        volumeMounts:
        - mountPath: /mnt/
          name: config-volume
          readOnly: false
        - mountPath: /data/
          name: redis-cluster-storage
          readOnly: false
      serviceAccountName: redis-cluster
      terminationGracePeriodSeconds: 30
      volumes:
      - configMap:
```

```
            items:
            - key: redis-cluster.conf
              path: redis-cluster.conf
            name: redis-cluster-config
          name: config-volume
  volumeClaimTemplates:
  - metadata:
      name: redis-cluster-storage
    spec:
      accessModes:
      - ReadWriteOnce
      storageClassName: "redis-cluster-storage-class"
      resources:
        requests:
          storage: 1Gi
```

此文件用于创建 Redis 实例，不一定非要使用 StatefulSet，采用 Deployment 部署也可。本例创建 6 个实例，用于集群中节点的 3 主 3 从。

3.3.2 创建 Redis 命名空间

可以将公用服务都放置在同一个 Namespace（命名空间）下，比如 public-service 中，然后按需修改。

创建 Namespace：

```
kubectl create namespace public-service
```

如果需要部署到其他 Namespace，需要更改当前目录中所有文件的 namespace：

```
sed -i "s#public-service#YOUR_NAMESPACE#g" *
```

3.3.3 创建 Redis 集群 PV

动态 PV 无须此步骤，本节使用的是 NFS 作为 PV，在实际使用中，Redis 的数据不一定需要做持久化，按需配置即可。

配置 NFS 服务器：

```
NFS 服务器：
    /K8S/redis-cluster/1 *(rw,sync,no_subtree_check,no_root_squash)
    /K8S/redis-cluster/2 *(rw,sync,no_subtree_check,no_root_squash)
    /K8S/redis-cluster/3 *(rw,sync,no_subtree_check,no_root_squash)
    /K8S/redis-cluster/4 *(rw,sync,no_subtree_check,no_root_squash)
    /K8S/redis-cluster/5 *(rw,sync,no_subtree_check,no_root_squash)
    /K8S/redis-cluster/6 *(rw,sync,no_subtree_check,no_root_squash)
```

此目录需要和 PV 文件的 path 一致。

3.3.4 创建集群

本节文件都在 chap03/3.3 目录下，请执行以下命令安装：

```
[root@K8S-master01 3.3]# ls
failover.py    redis-cluster-configmap.yaml    redis-cluster-rbac.yaml
redis-cluster-ss.yaml
README.md    redis-cluster-pv.yaml    redis-cluster-service.yaml
[root@K8S-master01 redis-cluster]# kubectl apply -f .
```

3.3.5 创建 slot

Redis 集群模式和 Redis 哨兵模式有所不同,等待节点全部启动后,开始创建 slot:

```
[root@K8S-master01 3.3]#v=""
[root@K8S-master01 3.3]# for i in `kubectl get po -n public-service -o wide | grep redis-cluster | awk '{print $6}' | grep -v IP`; do v="$v $i:6379";done
[root@K8S-master01 3.3]# kubectl exec -ti redis-cluster-ss-5 -n public-service -- redis-trib.rb create --replicas 1 $v
```

查看集群状态:

```
[root@K8S-master01 ~]# kubectl exec -ti redis-cluster-ss-0 -n public-service -- redis-cli cluster nodes
f9527e2ced3c472caabe3f815d87531e82e75049 172.168.5.174:6379@16379 master - 0 1541693210490 2 connected 5461-10922
a47ef989862a2ddbf83c70d8191ff17c8b37a6fc 172.168.2.68:6379@16379 master - 0 1541693213497 3 connected 10923-16383
b4c3d1ffe5ed70d2d40467d228004f4e0fb5fa25 172.168.5.175:6379@16379 slave f9527e2ced3c472caabe3f815d87531e82e75049 0 1541693216510 6 connected
2aa4d2e5de3aca325bff95325102da72334a5164 172.168.1.76:6379@16379 master - 0 1541693214503 7 connected 0-5460
74c6e2356e41c6842e05b043c48ce20b7f1ad3ae 172.168.0.95:6379@16379 slave a47ef989862a2ddbf83c70d8191ff17c8b37a6fc 0 1541693215504 4 connected
2d5389ff7ff6b6dcc5cff83654a6e15c9c4a7750 172.168.6.170:6379@16379 myself,slave 2aa4d2e5de3aca325bff95325102da72334a5164 0 1541692127160 1 connected
[root@K8S-master01 ~]# kubectl exec -ti redis-cluster-ss-0 -n public-service -- redis-cli cluster info
    cluster_state:ok
    cluster_slots_assigned:16384
    cluster_slots_ok:16384
    cluster_slots_pfail:0
    cluster_slots_fail:0
    cluster_known_nodes:6
    cluster_size:3
    cluster_current_epoch:7
    cluster_my_epoch:7
    cluster_stats_messages_ping_sent:1158
    cluster_stats_messages_pong_sent:1203
    cluster_stats_messages_sent:2361
    cluster_stats_messages_ping_received:1203
    cluster_stats_messages_pong_received:1142
    cluster_stats_messages_received:2345
```

> **注 意**
>
> Redis 各实例最好不要部署在同一个节点上,以免节点故障,因为单个 Redis 故障并不会影响集群的稳定性。

3.4 安装 RabbitMQ 集群到 K8S 集群中

本例安装仍采用 NFS 作为后端存储，同样也可以采用 GFS 作为动态存储，更改方式和上一节类似，本次安装也部署在 public-service 命名空间内。和 Redis 集群类似，持久化部署也不是必须的，同样也可以采用 hostPath 进行挂载等。

3.4.1 各文件解释

1. rabbitmq-configmap.yaml

同样将 RabbitMQ 的配置文件放置到 ConfigMap 中，定义 ConfigMap 的文件 rabbitmq-configmap.yaml 如下（需要开启的插件请按需修改）：

```yaml
kind: ConfigMap
apiVersion: v1
metadata:
  name: rmq-cluster-config
  namespace: public-service
  labels:
    addonmanager.kubernetes.io/mode: Reconcile
data:
  enabled_plugins: |
    [rabbitmq_management,rabbitmq_peer_discovery_K8S].
  rabbitmq.conf: |
    loopback_users.guest = false
    ## Clustering
    cluster_formation.peer_discovery_backend = rabbit_peer_discovery_K8S
    cluster_formation.K8S.host = kubernetes.default.svc.cluster.local
    cluster_formation.K8S.address_type = hostname
    ####################################################
    # public-service is rabbitmq-cluster's namespace#
    ####################################################
    cluster_formation.K8S.hostname_suffix = .rmq-cluster.public-service.svc.cluster.local
    cluster_formation.node_cleanup.interval = 10
    cluster_formation.node_cleanup.only_log_warning = true
    cluster_partition_handling = autoheal
    ## queue master locator
    queue_master_locator=min-masters
```

2. rabbitmq-pv.yaml

定义 RabbitMQ 的持久化文件，即 rabbitmq-pv.yaml：

```yaml
apiVersion: v1
kind: PersistentVolume
metadata:
  name: pv-rmq-1
spec:
```

```yaml
  capacity:
    storage: 4Gi
  accessModes:
    - ReadWriteMany
  volumeMode: Filesystem
  persistentVolumeReclaimPolicy: Recycle
  storageClassName: "rmq-storage-class"
  nfs:
    # real share directory
    path: /K8S/rmq-cluster/rabbitmq-cluster-1
    # nfs real ip
    server: 192.168.2.2

---
apiVersion: v1
kind: PersistentVolume
metadata:
  name: pv-rmq-2
spec:
  capacity:
    storage: 4Gi
  accessModes:
    - ReadWriteMany
  volumeMode: Filesystem
  persistentVolumeReclaimPolicy: Recycle
  storageClassName: "rmq-storage-class"
  nfs:
    # real share directory
    path: /K8S/rmq-cluster/rabbitmq-cluster-2
    # nfs real ip
    server: 192.168.2.2

---
apiVersion: v1
kind: PersistentVolume
metadata:
  name: pv-rmq-3
spec:
  capacity:
    storage: 4Gi
  accessModes:
    - ReadWriteMany
  volumeMode: Filesystem
  persistentVolumeReclaimPolicy: Recycle
  storageClassName: "rmq-storage-class"
  nfs:
    # real share directory
    path: /K8S/rmq-cluster/rabbitmq-cluster-3
    # nfs real ip
    server: 192.168.2.2
```

本例 RabbitMQ 持久化所用的也是静态 PV，动态 PV 无须创建此文件，此处使用的是 NFS 作为后端存储，请按需修改。

3. rabbitmq-cluster-ss.yaml

本例安装同样采用 StatefulSet 的形式，定义 StatefulSet 文件 rabbitmq-cluster-ss.yaml 如下：

```yaml
kind: StatefulSet
apiVersion: apps/v1beta1
metadata:
  labels:
    app: rmq-cluster
  name: rmq-cluster
  namespace: public-service
spec:
  replicas: 3
  selector:
    matchLabels:
      app: rmq-cluster
  serviceName: rmq-cluster
  template:
    metadata:
      labels:
        app: rmq-cluster
    spec:
      containers:
      - args:
        - -c
        - cp -v /etc/rabbitmq/rabbitmq.conf ${RABBITMQ_CONFIG_FILE}; exec docker-entrypoint.sh
          rabbitmq-server
        command:
        - sh
        env:
        - name: RABBITMQ_DEFAULT_USER
          valueFrom:
            secretKeyRef:
              key: username
              name: rmq-cluster-secret
        - name: RABBITMQ_DEFAULT_PASS
          valueFrom:
            secretKeyRef:
              key: password
              name: rmq-cluster-secret
        - name: RABBITMQ_ERLANG_COOKIE
          valueFrom:
            secretKeyRef:
              key: cookie
              name: rmq-cluster-secret
        - name: K8S_SERVICE_NAME
          value: rmq-cluster
        - name: POD_IP
          valueFrom:
            fieldRef:
              fieldPath: status.podIP
        - name: POD_NAME
          valueFrom:
            fieldRef:
```

```yaml
            fieldPath: metadata.name
      - name: POD_NAMESPACE
        valueFrom:
          fieldRef:
            fieldPath: metadata.namespace
      - name: RABBITMQ_USE_LONGNAME
        value: "true"
      - name: RABBITMQ_NODENAME
        value: rabbit@$(POD_NAME).rmq-cluster.$(POD_NAMESPACE).svc.cluster.local
      - name: RABBITMQ_CONFIG_FILE
        value: /var/lib/rabbitmq/rabbitmq.conf
      image: rabbitmq:3.7-management
      imagePullPolicy: IfNotPresent
      livenessProbe:
        exec:
          command:
          - rabbitmqctl
          - status
        initialDelaySeconds: 30
        timeoutSeconds: 10
      name: rabbitmq
      ports:
      - containerPort: 15672
        name: http
        protocol: TCP
      - containerPort: 5672
        name: amqp
        protocol: TCP
      readinessProbe:
        exec:
          command:
          - rabbitmqctl
          - status
        initialDelaySeconds: 10
        timeoutSeconds: 10
      volumeMounts:
      - mountPath: /etc/rabbitmq
        name: config-volume
        readOnly: false
      - mountPath: /var/lib/rabbitmq
        name: rabbitmq-storage
        readOnly: false
    serviceAccountName: rmq-cluster
    terminationGracePeriodSeconds: 30
    volumes:
    - configMap:
        items:
        - key: rabbitmq.conf
          path: rabbitmq.conf
        - key: enabled_plugins
          path: enabled_plugins
        name: rmq-cluster-config
      name: config-volume
  volumeClaimTemplates:
```

```
    - metadata:
        name: rabbitmq-storage
      spec:
        accessModes:
        - ReadWriteMany
        storageClassName: "rmq-storage-class"
        resources:
          requests:
            storage: 4Gi
```

如果使用 GFS 动态存储需要更改 storageClassName 的值。

3.4.2 配置 NFS

此处采用的是静态 PV 方式，后端使用的是 NFS，为了方便扩展可以使用动态 PV：

```
[root@nfs rabbitmq-cluster-1]# cat /etc/exports
/K8S/rmq-cluster/rabbitmq-cluster-1/ *(rw,sync,no_subtree_check,
no_root_squash)
/K8S/rmq-cluster/rabbitmq-cluster-2/ *(rw,sync,no_subtree_check,
no_root_squash)
/K8S/rmq-cluster/rabbitmq-cluster-3/ *(rw,sync,no_subtree_check,
no_root_squash)
```

3.4.3 创建集群

这里采用动态存储创建集群，这样就无须创建 rabbitmq-pv.yaml 文件。

如需更改 Namespace，要和部署 Redis 时的 Namespace 一致：

```
[root@K8S-master01 rabbitmq-cluster]# pwd
/root/efk/rabbitmq-cluster
[root@K8S-master01 rabbitmq-cluster]# ls
rabbitmq-cluster-ss.yaml  rabbitmq-pv.yaml   rabbitmq-secret.yaml
rabbitmq-service-lb.yaml
rabbitmq-configmap.yaml   rabbitmq-rbac.yaml  rabbitmq-service-cluster.yaml
README.md

[root@K8S-master01 rabbitmq-cluster]# kubectl apply -f .
```

3.4.4 查看资源

查看集群中的各类资源：

```
[root@K8S-master01 rabbitmq-cluster]# kubectl get pods -n public-service
NAME             READY   STATUS    RESTARTS   AGE
rmq-cluster-0    1/1     Running   0          40m
rmq-cluster-1    1/1     Running   0          39m
rmq-cluster-2    1/1     Running   0          39m


[root@K8S-master01 rabbitmq-cluster]# kubectl get pv -n public-service
```

```
        NAME           CAPACITY    ACCESS MODES    RECLAIM POLICY    STATUS     CLAIM
STORAGECLASS          REASON      AGE
        pv-rmq-1       4Gi         RWX             Recycle           Bound
public-service/rabbitmq-storage-rmq-cluster-2        rmq-storage-class               49m
        pv-rmq-2       4Gi         RWX             Recycle           Bound
public-service/rabbitmq-storage-rmq-cluster-1        rmq-storage-class               49m
        pv-rmq-3       4Gi         RWX             Recycle           Bound
public-service/rabbitmq-storage-rmq-cluster-0        rmq-storage-class               49m
        [root@K8S-master01 rabbitmq-cluster]# kubectl get pvc -n public-service
        NAME                              STATUS    VOLUME      CAPACITY    ACCESS MODES
STORAGECLASS          AGE
        rabbitmq-storage-rmq-cluster-0    Bound     pv-rmq-3    4Gi         RWX
rmq-storage-class     48m
        rabbitmq-storage-rmq-cluster-1    Bound     pv-rmq-2    4Gi         RWX
rmq-storage-class     48m
        rabbitmq-storage-rmq-cluster-2    Bound     pv-rmq-1    4Gi         RWX
rmq-storage-class     48m
        [root@K8S-master01 rabbitmq-cluster]#
```

查看 Service，此时使用的 NodePort 方式可改为 Ingress：

```
        [root@K8S-master01 rabbitmq-cluster]# kubectl get services -n public-service
        NAME                   TYPE         CLUSTER-IP      EXTERNAL-IP    PORT(S)
AGE
        rmq-cluster            ClusterIP    None            <none>         5672/TCP
1h
        rmq-cluster-balancer   NodePort     10.107.221.85   <none>
15672:30051/TCP,5672:31892/TCP    1h
```

3.4.5 访问测试

通过 NodePort 的端口进行访问，账号密码 guest，参考图 3-1。

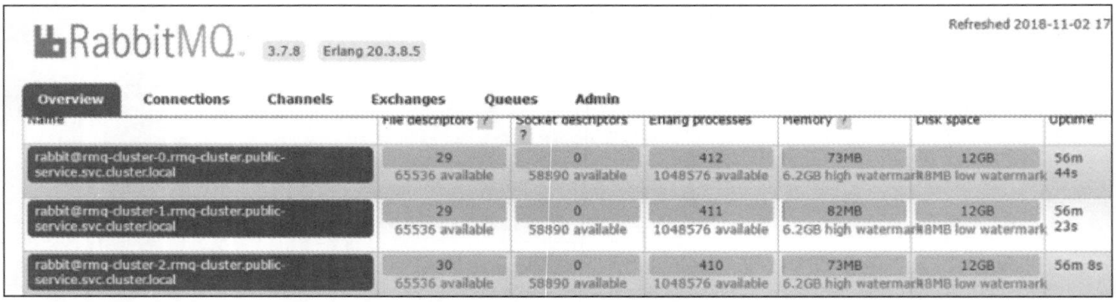

图 3-1 RabbitMQ 页面

3.5 安装 GitLab 到 K8S 集群中

本节将持久化安装 GitLab 到 K8S 集群中，但在生产环境中还是建议将 GitLab 部署至集群之外。本例安装的是 GitLab-ce，采用 GFS 实现动态存储（并非必须的，本书仅作为演示而选用了

GFS），GitLab-ce 所用数据库为 PostgreSQL，采用 NFS 存储。

3.5.1 各文件介绍

1. gitlab-pvc.yaml

创建 GitLab 的持久化文件，此文件定义的是 GitLab 使用 GFS 作为动态存储，其他存储方式定义也类似，只是 storageClassName 不一致。这里定义 PVC 文件 gitlab-pvc.yaml 如下：

```yaml
kind: PersistentVolumeClaim
apiVersion: v1
metadata:
  name: gitlab-gitlab
  namespace: public-service
spec:
  accessModes: [ "ReadWriteMany" ]
  storageClassName: "gluster-heketi"
  resources:
    requests:
      storage: 5Gi
```

2. gitlab-rc.yml

本例安装直接使用 ReplicationController 进行安装（官方建议使用 helm 进行安装，安装的是 GitLab 全家桶，包括 Git Runner，因为无须其他功能，所以只部署 GitLab），定义 gitlab-rc.yml 文件如下：

```yaml
apiVersion: v1
kind: ReplicationController
metadata:
  name: gitlab-ldap
  namespace: public-service
spec:
  replicas: 1
  selector:
    name: gitlab
  template:
    metadata:
      name: gitlab
      labels:
        name: gitlab
    spec:
      containers:
      - name: gitlab
        image: sameersbn/gitlab:11.5.1
        env:
        - name: TZ
          value: Asia/Shanghai
        - name: GITLAB_TIMEZONE
          value: Beijing

        - name: GITLAB_SECRETS_DB_KEY_BASE
          value: long-and-random-alpha-numeric-string
```

```yaml
        - name: GITLAB_SECRETS_SECRET_KEY_BASE
          value: long-and-random-alpha-numeric-string
        - name: GITLAB_SECRETS_OTP_KEY_BASE
          value: long-and-random-alpha-numeric-string

        - name: GITLAB_ROOT_PASSWORD
          value: gitlab123
        - name: GITLAB_ROOT_EMAIL
          value: gitlab@xxx.com

        - name: GITLAB_HOST
          value: gitlab.xxx.net
        - name: GITLAB_PORT
          value: "80"
        - name: GITLAB_SSH_PORT
          value: "22"

        - name: GITLAB_NOTIFY_ON_BROKEN_BUILDS
          value: "true"
        - name: GITLAB_NOTIFY_PUSHER
          value: "false"

        - name: GITLAB_BACKUP_SCHEDULE
          value: daily
        - name: GITLAB_BACKUP_TIME
          value: 01:00

        - name: DB_TYPE
          value: postgres
        - name: DB_HOST
          value: postgresql
        - name: DB_PORT
          value: "5432"
        - name: DB_USER
          value: gitlab
        - name: DB_PASS
          value: passw0rd
        - name: DB_NAME
          value: gitlab_production

        - name: REDIS_HOST
          value: redis
        - name: REDIS_PORT
          value: "6379"

        - name: SMTP_ENABLED
          value: "true"
        - name: SMTP_DOMAIN
          value: smtp.exmail.qq.com
        - name: SMTP_HOST
          value: smtp.exmail.qq.com
        - name: SMTP_PORT
          value: "465"
        - name: SMTP_USER
          value: dukuan@xxx.com
```

```yaml
        - name: SMTP_PASS
          value: "DKxxx"
        - name: SMTP_STARTTLS
          value: "true"
        - name: SMTP_AUTHENTICATION
          value: login

        - name: IMAP_ENABLED
          value: "false"
        - name: IMAP_HOST
          value: imap.gmail.com
        - name: IMAP_PORT
          value: "993"
        - name: IMAP_USER
          value: mailer@example.com
        - name: IMAP_PASS
          value: password
        - name: IMAP_SSL
          value: "true"
        - name: IMAP_STARTTLS
          value: "false"
        - name: LDAP_ENABLED
          value: "true"
        - name: LDAP_LABEL
          value: 'LDAP'
        - name: LDAP_HOST
          value: 'ldap-service'
        - name: LDAP_PORT
          value: '389'
        - name: LDAP_UID
          value: 'uid'
        - name: LDAP_BIND_DN
          value: 'cn=admin,dc=example,dc=org'
        - name: LDAP_PASS
          value: 'admin'
        - name: LDAP_BASE
          value: 'dc=example,dc=org'
        - name: LDAP_ALLOW_USERNAME_OR_EMAIL_LOGIN
          value: "true"
        - name: LDAP_VERIFY_SSL
          value: 'false'
        - name: LDAP_METHOD
          value: 'plain'

        ports:
        - name: http
          containerPort: 80
        - name: ssh
          containerPort: 22
        volumeMounts:
        - mountPath: /home/git/data
          name: data
        livenessProbe:
          httpGet:
            path: /
```

```yaml
          port: 80
          initialDelaySeconds: 180
          timeoutSeconds: 5
        readinessProbe:
          httpGet:
            path: /
            port: 80
          initialDelaySeconds: 5
          timeoutSeconds: 1
      volumes:
      - name: data
        persistentVolumeClaim:
          claimName: gitlab-gitlab
```

其中，claimName 为上述 gitlab-pvc.yaml 创建的 PVC 的名字。

需要更改的地方有 Redis 和 PostgreSQL 配置（如果是内置的，则无须更改），SMTP 和 IMAP 配置，LDAP 认证登录（不采用 LDAP 认证，无须配置，删除即可）。

3. pg-pv.yaml

定义 PostgreSQL 的持久化文件 pg-pv.yaml 如下：

```yaml
apiVersion: v1
kind: PersistentVolume
metadata:
  name: gitlab-pg-data
spec:
  capacity:
    storage: 5Gi
  accessModes:
    - ReadWriteMany
  volumeMode: Filesystem
  persistentVolumeReclaimPolicy: Recycle
  storageClassName: "gitlab-pg-data"
  nfs:
    # real share directory
    path: /K8S/gitlab-pg
    # nfs real ip
    server: 192.168.2.2
```

4. traefik-gitlab.yml

定义 GitLab 的 Ingress 文件 traefik-gitlab.yml 如下。如果使用 Nginx 作为 Ingress，只需要将 kubernetes.io/ingress.class 改为 Nginx 即可，域名按需修改：

```yaml
apiVersion: extensions/v1beta1
kind: Ingress
metadata:
  name: gitlab
  namespace: public-service
  annotations:
    kubernetes.io/ingress.class: traefik
spec:
  rules:
  - host: gitlab.xxx.net
```

```
    http:
      paths:
      - backend:
          serviceName: gitlab
          servicePort: 80
```

3.5.2 创建 GitLab

更改完配置后直接创建 GitLab：

```
[root@K8S-master01 gitlab]# kubectl apply -f .
```

查看资源：

```
[root@K8S-master01 gitlab]# kubectl get po,svc,pvc -n public-service
NAME                      READY     STATUS    RESTARTS   AGE
pod/gitlab-cctr6          1/1       Running   2          37m
pod/postgresql-c6trh      1/1       Running   1          37m
pod/redis-b6vfk           1/1       Running   0          3h

NAME                                            TYPE        CLUSTER-IP       EXTERNAL-IP
PORT(S)                       AGE
    service/gitlab                              ClusterIP   10.109.163.143   <none>
80/TCP,22/TCP                 24m
    service/gitlab-balancer                     NodePort    10.108.77.162    <none>
80:30049/TCP,22:30347/TCP     14m
    service/glusterfs-dynamic-gitlab-gitlab     ClusterIP   10.102.192.68    <none>
1/TCP                         59m
    service/glusterfs-dynamic-gitlab-pg         ClusterIP   10.96.14.147     <none>
1/TCP                         37m
    service/glusterfs-dynamic-gitlab-redis      ClusterIP   10.106.253.41    <none>
1/TCP                         1h
    service/postgresql                          ClusterIP   10.104.102.20    <none>
5432/TCP                      3h
    service/redis                               ClusterIP   10.97.174.50     <none>
6379/TCP                      3h

    NAME                                            STATUS    VOLUME
CAPACITY   ACCESS MODES   STORAGECLASS    AGE
    persistentvolumeclaim/gitlab-gitlab         Bound
pvc-b8249829-f6bf-11e8-9640-000c298bf023   5Gi        RWX
gluster-heketi   59m
    persistentvolumeclaim/gitlab-pg             Bound
pvc-b40b6227-f6c2-11e8-9640-000c298bf023   5Gi        RWX
gluster-heketi   37m
    persistentvolumeclaim/gitlab-redis          Bound
pvc-28d0276d-f6af-11e8-8d2c-000c293bfe27   3Gi        RWX
gluster-heketi   2h
```

3.5.3 访问 GitLab

URL 为 Ingress 定义的 URL，默认账号，密码是 gitlab123，参考图 3-2。

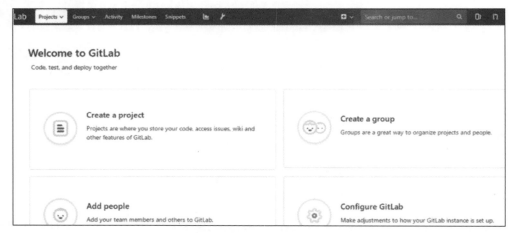

图 3-2　GitLab 页面

语言更改。注意，此时翻译是实验性的，参考图 3-3，选择 Setting 页面。

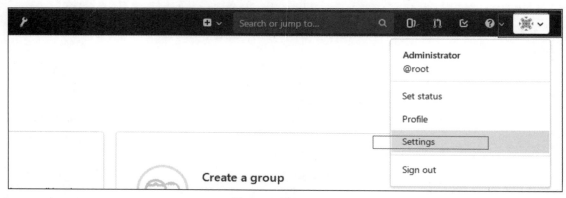

图 3-3　选择 Settings

之后选择简体中文，如图 3-4 所示。

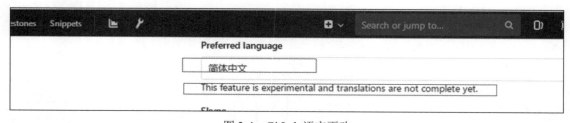

图 3-4　GitLab 语言更改

3.5.4　创建项目

首先创建组，如图 3-5 所示。

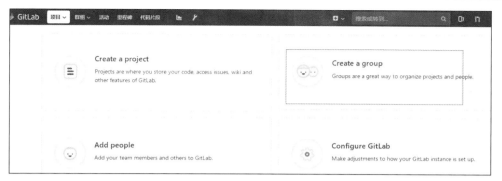

图 3-5　创建 Group

输入 Group 信息，如图 3-6 所示。

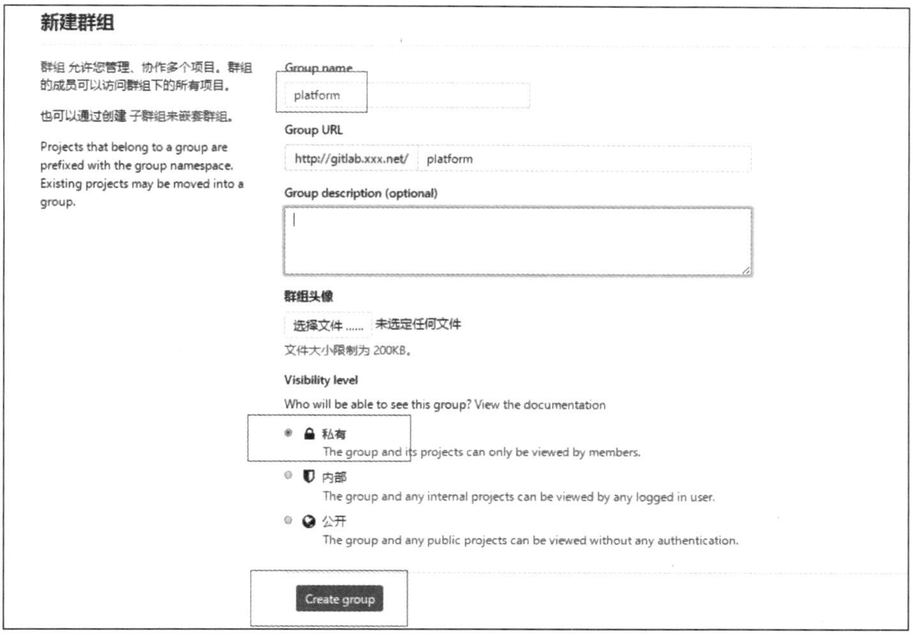

图 3-6　Group 信息输入

创建项目，如图 3-7 所示。

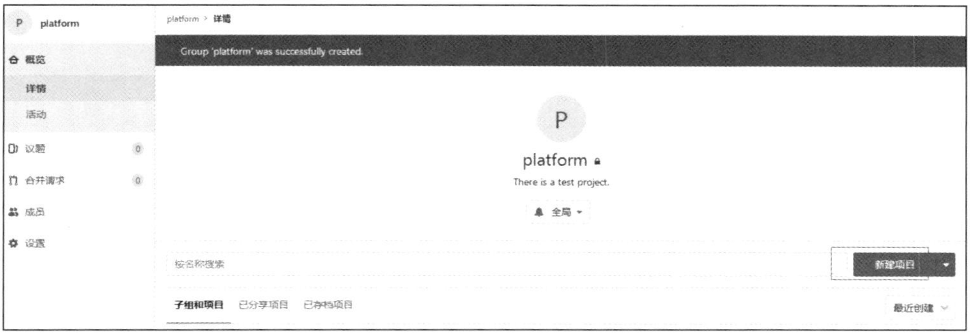

图 3-7　创建项目

编辑项目信息，如图 3-8 所示。

图 3-8　项目配置

添加 README，如图 3-9 所示。

图 3-9　添加 README

3.5.5　创建用户权限

添加权限，一般用于区分开发者用户或者管理员角色，如图 3-10 所示。

图 3-10　添加权限

配置用户权限，如图 3-11 所示。

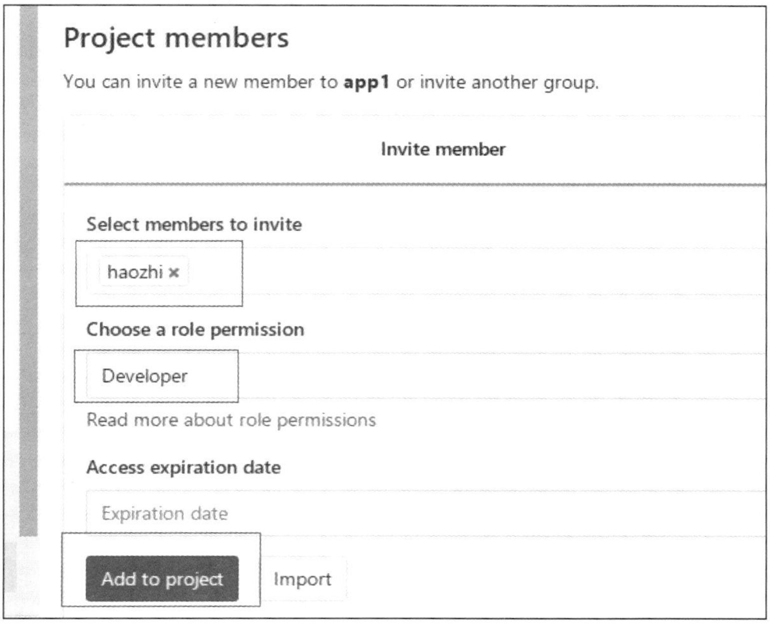

图 3-11　配置用户权限

登录此用户即可查看此项目，如图 3-12 所示。

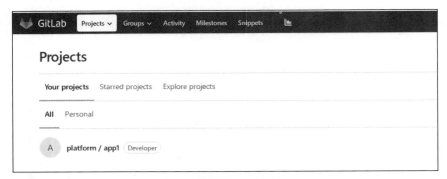

图 3-12 登录其他用户

3.5.6 添加 SSH Key

为工作人员添加 SSH Key,可以实现无密码访问,如图 3-13 所示。

图 3-13 配置 SSH Key

添加用户 Key 至 GitLab,如图 3-14 所示。

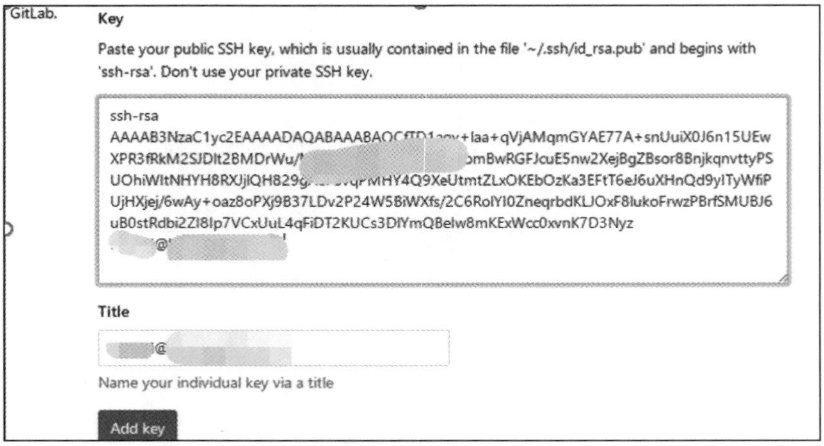

图 3-14 添加用户 Key

配置完成页面如图 1-15 所示。

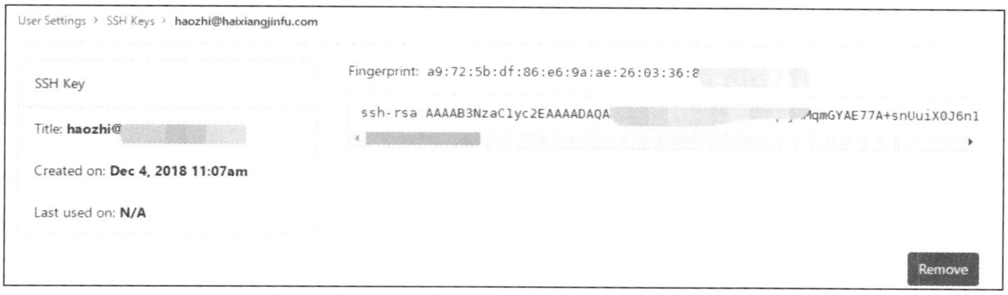

第 3 章　Kubernetes 常见应用安装 | 179

图 3-15　完成添加 Key

如果没有 Key，则需要使用 ssh-keygen -t rsa -C "your@email.com" 生成对应的 Key。

3.5.7　项目开发

首先根据图 3-16 在自己的工作台完成基本配置。

图 3-16　基本配置

创建分支。项目开发时，一般新建分支进行开发，上线无误后再合并到主分支，如图 3-17 所示。

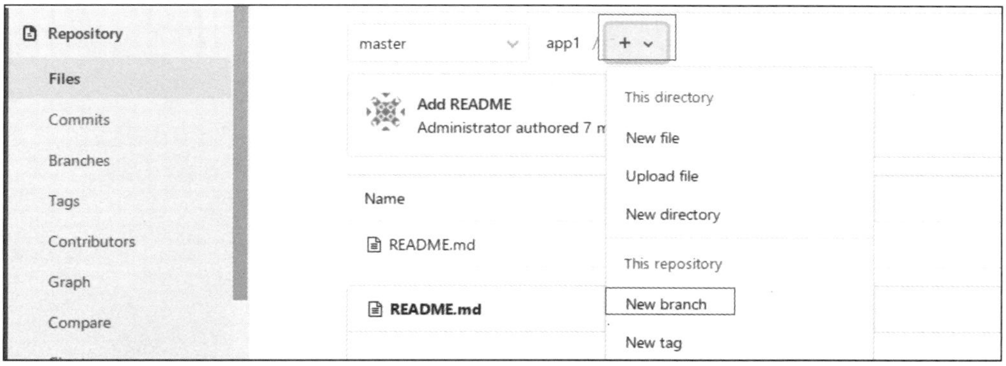

图 3-17　选择添加新分支

输入对应的分支名，按需更改分支名（Branch name），如图 3-18 所示。

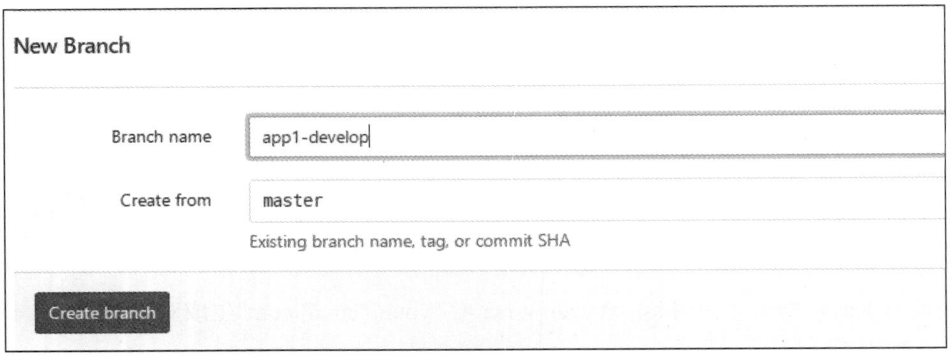

图 3-18　配置分支名

克隆代码。注意，此时需要更改 Git 的地址，因为 SSH 端口并非 22，可以通过 Service 查看 NodePort 的端口：

```
λ git clone ssh://git@gitlab.xxx.net:32455/platform/app1.git
Cloning into 'app1'...
remote: Enumerating objects: 3, done.
remote: Counting objects: 100% (3/3), done.
remote: Total 3 (delta 0), reused 0 (delta 0)
Receiving objects: 100% (3/3), done.
Checking connectivity... done.

D:\code
λ cd app1\

D:\code\app1 (master)
λ git branch -a
* master
  remotes/origin/HEAD -> origin/master
  remotes/origin/app1-develop
  remotes/origin/master

D:\code\app1 (master)
λ git checkout app1-develop
Branch app1-develop set up to track remote branch app1-develop from origin.
Switched to a new branch 'app1-develop'

D:\code\app1 (app1-develop)
λ touch.exe testfile

D:\code\app1 (app1-develop)
λ git add .

D:\code\app1 (app1-develop)
λ git commit -am "create a test file"
[app1-develop 9050e35] create a test file
 1 file changed, 0 insertions(+), 0 deletions(-)
 create mode 100644 testfile
```

```
D:\code\app1 (app1-develop)
λ git push origin app1-develop
Counting objects: 3, done.
Delta compression using up to 4 threads.
Compressing objects: 100% (2/2), done.
Writing objects: 100% (3/3), 278 bytes | 0 bytes/s, done.
Total 3 (delta 0), reused 0 (delta 0)
remote:
remote: To create a merge request for app1-develop, visit:
remote:
http://gitlab.xxx.net/platform/app1/merge_requests/new?merge_request%5Bsource_branch%5D=app1-develop remote:
   To ssh://git@gitlab.xxx.net:32455/platform/app1.git
    0a63d86..9050e35  app1-develop -> app1-develop
```

查看刚才提交的文件，如图 3-19 所示。

图 3-19　查看 Commit 的文件

协同开发。以同样方式将其他用户加入此项目。

克隆代码并修改文件：

```
[root@K8S-node02 ~]# git clone ssh://git@gitlab.xxx.net:32455/platform/app1.git
Cloning into 'app1'...
The authenticity of host '[gitlab.xxx.net]:32455 ([192.168.20.10]:32455)' can't be established.
ECDSA key fingerprint is SHA256:l6BYlMWpAWyXx/f5oTG8lK4JQvG9C2ZZ9opqdQZfIuc.
ECDSA key fingerprint is MD5:5b:b4:04:68:26:53:2e:ba:fe:f8:99:6c:8f:d3:fa:51.
Are you sure you want to continue connecting (yes/no)? yes
Warning: Permanently added '[gitlab.xxx.net]:32455,[192.168.20.10]:32455' (ECDSA) to the list of known hosts.
remote: Enumerating objects: 6, done.
remote: Counting objects: 100% (6/6), done.
remote: Compressing objects: 100% (3/3), done.
remote: Total 6 (delta 0), reused 0 (delta 0)
Receiving objects: 100% (6/6), done.
[root@K8S-node02 ~]# cd app1/
```

```
[root@K8S-node02 app1]# ls
README.md
[root@K8S-node02 app1]# git branch -a
* master
  remotes/origin/HEAD -> origin/master
  remotes/origin/app1-develop
  remotes/origin/master
[root@K8S-node02 app1]# git checkout app1-develop
Branch app1-develop set up to track remote branch app1-develop from origin.
Switched to a new branch 'app1-develop'
[root@K8S-node02 app1]# ls
README.md  testfile
[root@K8S-node02 app1]# echo "add something" >> testfile
[root@K8S-node02 app1]# git add .
[root@K8S-node02 app1]# git commit -am "add someting to testfile"
[app1-develop 69d693c] add someting to testfile
 1 file changed, 1 insertion(+)
[root@K8S-node02 app1]# git push origin app1-develop
Counting objects: 5, done.
Delta compression using up to 4 threads.
Compressing objects: 100% (2/2), done.
Writing objects: 100% (3/3), 305 bytes | 0 bytes/s, done.
Total 3 (delta 0), reused 0 (delta 0)
remote:
remote: To create a merge request for app1-develop, visit:
remote:
http://gitlab.xxx.net/platform/app1/merge_requests/new?merge_request%5Bsource_branch%5D=app1-develop
remote:
To ssh://git@gitlab.xxx.net:32455/platform/app1.git
   9050e35..69d693c  app1-develop -> app1-develop
```

3.6 安装 Jenkins 到 K8S 集群中

Jenkins 在 DevOps 工具链中是核心的流程管理中心，可用来复制串联系统的构建流程、测试流程、镜像制作流程、部署流程等。本节首先使用 Helm 安装 Jenkins，关于 Jenkins 的使用在持续集成与持续部署一章中会详细讲解。一般流程是通过 Redmine 和 GitLab 创建项目、开发提交代码、触发 Jenkins 自动构建并生成镜像（image）上传至私有镜像仓库，然后自动部署到 K8S 集群。

3.6.1 各文件介绍

本例安装 Jenkins 采用的方式是 Helm，之前安装的应用都可以采用 Helm 进行安装。以下按需修改 values.yaml 对应的文件：

```
Master:
  Name: jenkins-master
  Image: "jenkins/jenkins"
  ImageTag: "lts"
  ImagePullPolicy: "Always"
```

```yaml
  Component: "jenkins-master"
  UseSecurity: true
  AdminUser: admin
  Cpu: "200m"
  Memory: "256Mi"
  ServicePort: 8080
  ServiceType: ClusterIP
  ServiceAnnotations: {}
  ContainerPort: 8080
  HealthProbes: true
  HealthProbesTimeout: 60
  SlaveListenerPort: 50000
  LoadBalancerSourceRanges:
  - 0.0.0.0/0
  InstallPlugins:
    - kubernetes:1.1
    - workflow-aggregator:2.5
    - workflow-job:2.15
    - credentials-binding:1.13
    - git:3.6.4
  InitScripts:
  CustomConfigMap: false
  NodeSelector: {}
  Tolerations: {}
  Ingress:
    Annotations:
    TLS:
Agent:
  Enabled: true
  Image: jenkins/jnlp-slave
  ImageTag: 3.10-1
  Component: "jenkins-slave"
  Privileged: false
  Cpu: "200m"
  Memory: "256Mi"
  AlwaysPullImage: false
  volumes:
  NodeSelector: {}
Persistence:
  Enabled: true
  StorageClass: "gluster-heketi"
  Annotations: {}
  AccessMode: ReadWriteMany
  Size: 20Gi
  volumes:
  mounts:
NetworkPolicy:
  Enabled: false
  ApiVersion: extensions/v1beta1
rbac:
  install: false
  serviceAccountName: default
  apiVersion: v1beta1
  roleRef: cluster-admin
```

主要修改的地方是 StorageClass，请选择持久化的方式。本例使用的是 GFS 作为动态存储，部署 Jenkins 最好进行持久化，不然 Jenkins 工具重启后，已安装的插件还需要重新下载，配置的任务也会丢失，所以持久化 Jenkins 是极为重要的一步。

其次，需要修改 Ingress 配置，也可不修改，本节代码提供了单独的 Ingress 配置，可直接使用。

3.6.2 安装 Jenkins

使用 Helm 安装 Jenkins：

```
[root@K8S-master01 3.6]# helm install --name jenkins . --namespace public-service
NAME:   jenkins
LAST DEPLOYED: Tue Dec  4 14:55:24 2018
NAMESPACE: public-service
STATUS: DEPLOYED

RESOURCES:
==> v1/Secret
NAME     AGE
jenkins  0s

==> v1/ConfigMap
jenkins         0s
jenkins-tests   0s

==> v1/PersistentVolumeClaim
jenkins  0s

==> v1/Service
jenkins-agent  0s
jenkins        0s

==> v1beta1/Deployment
jenkins  0s

==> v1/Pod(related)

NAME                       READY  STATUS   RESTARTS  AGE
jenkins-5b6c648956-zds2p   0/1    Pending  0         0s


NOTES:
1. Get your 'admin' user password by running:
   printf $(kubectl get secret --namespace public-service jenkins -o jsonpath="{.data.jenkins-admin-password}" | base64 --decode);echo
2. Get the Jenkins URL to visit by running these commands in the same shell:
   export POD_NAME=$(kubectl get pods --namespace public-service -l "component=jenkins-master" -o jsonpath="{.items[0].metadata.name}")
   echo http://127.0.0.1:8080
   kubectl port-forward $POD_NAME 8080:8080

3. Login with the password from step 1 and the username: admin
```

```
For more information on running Jenkins on Kubernetes, visit:
https://cloud.google.com/solutions/jenkins-on-container-engine
```

创建 Ingress：

```
[root@K8S-master01 jenkins]# kubectl create -f traefik-jenkins.yaml
ingress.extensions/jenkins created
```

查看状态：

```
[root@K8S-master01 ~]# kubectl get po,svc,ingress,pvc -n public-service | grep
jenkins
    pod/jenkins-5b6c648956-zds2p                   1/1    Running    5         44h
    service/glusterfs-dynamic-jenkins              ClusterIP   10.111.100.114
<none>       1/TCP                        44h
    service/jenkins                                ClusterIP   10.107.215.94
<none>       8080/TCP                     44h
    service/jenkins-agent                          ClusterIP   10.103.212.222
<none>       50000/TCP                    44h
    ingress.extensions/jenkins       jenkins.xxx.net              80    3m26s
    persistentvolumeclaim/jenkins                   Bound
pvc-953c3093-f791-11e8-9640-000c298bf023    20Gi       RWX
gluster-heketi-2    44h
```

3.6.3 访问 Jenkins

网址为 Ingress 定义的网址，如图 3-20 所示。

图 3-20　Jenkins 主页面

查看 Jenkins 管理员密码，账户为 values.yaml 里面设置的 admin：

```
[root@K8S-master01 ~]# kubectl get secret --namespace public-service jenkins
-o jsonpath="{.data.jenkins-admin-password}" | base64 --decode
9jni0dNNY9
```

登录：admin/9jni0dNNY9，如图 3-21 所示。

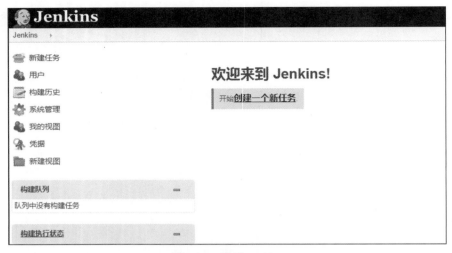

图 3-21 登录 Jenkins

至此 Jenkins 已经安装至 K8S，在实际使用时，Jenkins 也可以安装在集群之外。

3.7 安装 Harbor 到 K8S 集群中

本节安装 Harbor 到 K8S 集群中，Harbor 在集群中充当镜像仓库，持久化方式采用 GFS，如果采用其他方式，更改 values.yaml 和 redis-1.1.15.tgz 的 values.yaml 中的 storageClass 即可。其他配置比如 registry 的空间大小和域名按需修改即可，同样在生产环境中可将 Harbor 安装于集群之外。

3.7.1 安装 Harbor

使用 Helm 安装 Harbor：

```
helm install --name harbor-v1 . --wait --timeout 1500 --debug --namespace harbor
```

如果报出 forbidden 的错误，则需要创建 serviceaccount：

```
[root@K8S-master01 harbor-helm]# helm install --name harbor-v1 . --set externalDomain=harbor.xxx.net --wait --timeout 1500 --debug --namespace harbor
[debug] Created tunnel using local port: '35557'

[debug] SERVER: "127.0.0.1:35557"

[debug] Original chart version: ""
[debug] CHART PATH: /root/harbor-helm

Error: release harbor-v1 failed: namespaces "harbor" is forbidden: User "system:serviceaccount:kube-system:default" cannot get namespaces in the namespace "harbor"
```

```
kubectl create serviceaccount --namespace kube-system tiller

kubectl create clusterrolebinding tiller-cluster-rule
--clusterrole=cluster-admin --serviceaccount=kube-system:tiller

kubectl patch deploy --namespace kube-system tiller-deploy -p
'{"spec":{"template":{"spec":{"serviceAccount":"tiller"}}}}'
```

部署成功反馈如下：

```
......
==> v1/Pod(related)

NAME                                              READY   STATUS    RESTARTS   AGE
harbor-v1-redis-84dffd8574-xzrsh                  0/1     Running   0          <invalid>
harbor-v1-harbor-adminserver-5b59c684b4-g6cjc     1/1     Running   0          <invalid>
harbor-v1-harbor-chartmuseum-699cf6599-q6vfw      1/1     Running   0          <invalid>
harbor-v1-harbor-clair-6d9bb84485-2p52v           1/1     Running   0          <invalid>
harbor-v1-harbor-jobservice-5c9496775d-sj6mb      1/1     Running   0          <invalid>
harbor-v1-harbor-notary-server-5fb65b6866-dnnnk   1/1     Running   0          <invalid>
harbor-v1-harbor-notary-signer-5bfcfcd5cf-j774t   1/1     Running   0          <invalid>
harbor-v1-harbor-registry-75c9b6b457-pqxj6        1/1     Running   0          <invalid>
harbor-v1-harbor-ui-5974bd5549-zl9nj              1/1     Running   0          <invalid>
harbor-v1-harbor-database-0                       1/1     Running   0          <invalid>

==> v1/Secret

NAME                             AGE
harbor-v1-harbor-adminserver     <invalid>
harbor-v1-harbor-chartmuseum     <invalid>
harbor-v1-harbor-database        <invalid>
harbor-v1-harbor-ingress         <invalid>
harbor-v1-harbor-jobservice      <invalid>
harbor-v1-harbor-registry        <invalid>
harbor-v1-harbor-ui              <invalid>


NOTES:
Please wait for several minutes for Harbor deployment to complete.
Then you should be able to visit the UI portal at https://core.harbor.domain.
For more details, please visit https://github.com/goharbor/harbor.

......
```

查看资源：

```
[root@K8S-master01 harbor-helm]# kubectl get pod -n harbor
NAME                                                READY    STATUS     RESTARTS   AGE
harbor-v1-harbor-adminserver-5b59c684b4-g6cjc       1/1      Running    1          2m
harbor-v1-harbor-chartmuseum-699cf6599-q6vfw        1/1      Running    0          2m
harbor-v1-harbor-clair-6d9bb84485-2p52v             1/1      Running    1          2m
harbor-v1-harbor-database-0                         1/1      Running    0          2m
harbor-v1-harbor-jobservice-5c9496775d-sj6mb        1/1      Running    1          2m
harbor-v1-harbor-notary-server-5fb65b6866-dnnnk     1/1      Running    0          2m
harbor-v1-harbor-notary-signer-5bfcfcd5cf-j774t     1/1      Running    0          2m
harbor-v1-harbor-registry-75c9b6b457-pqxj6          1/1      Running    0          2m
harbor-v1-harbor-ui-5974bd5549-zl9nj                1/1      Running    2          2m
harbor-v1-redis-84dffd8574-xzrsh                    1/1      Running    0          2m

[root@K8S-master01 harbor-helm]# kubectl get svc -n harbor
NAME                                                                  TYPE        CLUSTER-IP        EXTERNAL-IP   PORT(S)     AGE
glusterfs-dynamic-database-data-harbor-v1-harbor-database-0           ClusterIP   10.101.10.82      <none>        1/TCP       2h
glusterfs-dynamic-harbor-v1-harbor-chartmuseum                        ClusterIP   10.97.114.51      <none>        1/TCP       36s
glusterfs-dynamic-harbor-v1-harbor-registry                           ClusterIP   10.98.207.16      <none>        1/TCP       36s
glusterfs-dynamic-harbor-v1-redis                                     ClusterIP   10.105.214.102    <none>        1/TCP       31s
harbor-v1-harbor-adminserver                                          ClusterIP   10.99.152.38      <none>        80/TCP      3m
harbor-v1-harbor-chartmuseum                                          ClusterIP   10.99.237.224     <none>        80/TCP      3m
harbor-v1-harbor-clair                                                ClusterIP   10.98.217.176     <none>        6060/TCP    3m
harbor-v1-harbor-database                                             ClusterIP   10.111.182.188    <none>        5432/TCP    3m
harbor-v1-harbor-jobservice                                           ClusterIP   10.98.202.61      <none>        80/TCP      3m
harbor-v1-harbor-notary-server                                        ClusterIP   10.110.72.98      <none>        4443/TCP    3m
harbor-v1-harbor-notary-signer                                        ClusterIP   10.106.234.19     <none>        7899/TCP    3m
harbor-v1-harbor-registry                                             ClusterIP   10.98.80.141      <none>        5000/TCP    3m
harbor-v1-harbor-ui                                                   ClusterIP   10.98.240.15      <none>        80/TCP      3m
harbor-v1-redis                                                       ClusterIP   10.107.234.107    <none>        6379/TCP    3m

[root@K8S-master01 harbor-helm]# kubectl get pv,pvc -n harbor | grep harbor
persistentvolume/pvc-080d1242-e990-11e8-8a89-000c293ad492   1Gi   RWO   Delete   Bound   harbor/database-data-harbor-v1-harbor-database-0   gluster-heketi   2h
persistentvolume/pvc-f573b165-e9a3-11e8-882f-000c293bfe27   8Gi   RWO   Delete   Bound   harbor/harbor-v1-redis                             gluster-heketi   1m
persistentvolume/pvc-f575855d-e9a3-11e8-882f-000c293bfe27   5Gi   RWO   Delete   Bound   harbor/harbor-v1-harbor-chartmuseum                gluster-heketi   1m
```

```
    persistentvolume/pvc-f577371b-e9a3-11e8-882f-000c293bfe27   10Gi         RWO
Delete         Bound      harbor/harbor-v1-harbor-registry
gluster-heketi                          1m
    persistentvolumeclaim/database-data-harbor-v1-harbor-database-0   Bound
pvc-080d1242-e990-11e8-8a89-000c293ad492    1Gi         RWO
gluster-heketi    2h
    persistentvolumeclaim/harbor-v1-harbor-chartmuseum               Bound
pvc-f575855d-e9a3-11e8-882f-000c293bfe27    5Gi         RWO
gluster-heketi    4m
    persistentvolumeclaim/harbor-v1-harbor-registry                  Bound
pvc-f577371b-e9a3-11e8-882f-000c293bfe27    10Gi        RWO
gluster-heketi    4m
    persistentvolumeclaim/harbor-v1-redis                            Bound
pvc-f573b165-e9a3-11e8-882f-000c293bfe27    8Gi         RWO
gluster-heketi    4m

[root@K8S-master01 harbor-helm]# vim values.yaml
[root@K8S-master01 harbor-helm]# kubectl get ingress -n harbor
    NAME                    HOSTS                              ADDRESS   PORTS
AGE
    harbor-v1-harbor-ingress   core.harbor.domain,notary.harbor.domain
80, 443    53m
```

3.7.2 访问 Harbor

访问测试。需要解析上述域名 core.harbor.domain 至 K8S 任意节点，如图 3-22 所示。

图 3-22 Harbor 主页面

默认账号密码：admin/Harbor12345，如图 3-23 所示。

图 3-23　Harbor 登录后页面

创建开发环境仓库，如图 3-24 所示。

图 3-24　选项新建项目

填写项目名称，如图 3-25 所示。

图 3-25　项目名称

3.7.3　在 K8S 中使用 Harbor

查看 Harbor 自带的证书，创建时被保存在 Secret 中：

```
[root@K8S-master01 ~]# kubectl get secrets/harbor-v1-harbor-ingress -n harbor
-o jsonpath="{.data.ca\.crt}" | base64 --decode
-----BEGIN CERTIFICATE-----
MIIC9DCCAdygAwIBAgIQffFj8E2+DLnbT3a3XRXlBjANBgkqhkiG9w0BAQsFADAU
MRIwEAYDVQQDEwloYXJib3ItY2EwHhcNMTgxMTE2MTYwODA5WhcNMjgxMTEzMTYw
ODA5WjAUMRIwEAYDVQQDEwloYXJib3ItY2EwggEiMA0GCSqGSIb3DQEBAQUAA4IB
DwAwggEKAoIBAQDw1WP6S3O+7zrhVAAZGcrAEdeQxr0c53eyDGcPL6my/h+FhZ1Y
KBvY5CLDVES957u/GtEXFfZr9aQT/PZECcccPcyZvt8NscEAuQONfrQFH/VLCvwm
XOcbFDR5BXDJR8nqGT6DVq8a1HUEOxiY39bp/Jz2HrDIfD9IMwEuyh/2IVXYHwD0
deaBpOY1slSylpOYWPFfy9UMfCsd+Jc7UCzRaiP3XWP9HMFKc4JTU8CDRR80s9UM
siU8QheVXn/Y9SxKaDfrYjaVUkEfJ6cAZkkDLmM1OzSU73N7I4nmm1SUS99vdSiZ
yu/R4oDFMezOkvYGBeDhLmmkK3sqWRh+dNoNAgMBAAGjQjBAMA4GA1UdDwEB/wQE
AwICpDAdBgNVHSUEFjAUBggrBgEFBQcDAQYIKwYBBQUHAwIwDwYDVR0TAQH/BAUw
AwEB/zANBgkqhkiG9w0BAQsFAAOCAQEAJjANauFSPZ+Da6VJSV2lGirpQN+EnrTl
u5VJxhQQGr1of4Je7aej6216KI9W5/Q4lDQfVOa/5JO1LFaiWp1AMBOlEm7FNiqx
LcLZzEZ4i6sLZ965FdrPGvy5cOeLa6D8Vx4faDCWaVYOkXoi/7oH91IuH6eEh+1H
u/Kelp8WEng4vfEcXRKkq4XTO51B1Mg1g7gflxMIoeSpXYSO5qwIL5ZqvoAD9H7J
CnQFO2xO3wrLq6TXH5Z7+0GWNghGk0GIOvF/ULHLWpsyhU5asKLK//MvORwQNHzL
b5LHG9uYeI+Jf12X4TI9qDaTCstiqM8vk1JPvgtSPJ9M62nRKY4ang==
-----END CERTIFICATE-----
```

在 Docker 工作目录下创建对应域名的证书：

```
cat <<EOF > /etc/docker/certs.d/core.harbor.domain/ca.crt
-----BEGIN CERTIFICATE-----
MIIC9DCCAdygAwIBAgIQffFj8E2+DLnbT3a3XRXlBjANBgkqhkiG9w0BAQsFADAU
MRIwEAYDVQQDEwloYXJib3ItY2EwHhcNMTgxMTE2MTYwODA5WhcNMjgxMTEzMTYw
ODA5WjAUMRIwEAYDVQQDEwloYXJib3ItY2EwggEiMA0GCSqGSIb3DQEBAQUAA4IB
DwAwggEKAoIBAQDw1WP6S3O+7zrhVAAZGcrAEdeQxr0c53eyDGcPL6my/h+FhZ1Y
KBvY5CLDVES957u/GtEXFfZr9aQT/PZECcccPcyZvt8NscEAuQONfrQFH/VLCvwm
XOcbFDR5BXDJR8nqGT6DVq8a1HUEOxiY39bp/Jz2HrDIfD9IMwEuyh/2IVXYHwD0
deaBpOY1slSylpOYWPFfy9UMfCsd+Jc7UCzRaiP3XWP9HMFKc4JTU8CDRR80s9UM
siU8QheVXn/Y9SxKaDfrYjaVUkEfJ6cAZkkDLmM1OzSU73N7I4nmm1SUS99vdSiZ
yu/R4oDFMezOkvYGBeDhLmmkK3sqWRh+dNoNAgMBAAGjQjBAMA4GA1UdDwEB/wQE
AwICpDAdBgNVHSUEFjAUBggrBgEFBQcDAQYIKwYBBQUHAwIwDwYDVR0TAQH/BAUw
AwEB/zANBgkqhkiG9w0BAQsFAAOCAQEAJjANauFSPZ+Da6VJSV2lGirpQN+EnrTl
u5VJxhQQGr1of4Je7aej6216KI9W5/Q4lDQfVOa/5JO1LFaiWp1AMBOlEm7FNiqx
LcLZzEZ4i6sLZ965FdrPGvy5cOeLa6D8Vx4faDCWaVYOkXoi/7oH91IuH6eEh+1H
u/Kelp8WEng4vfEcXRKkq4XTO51B1Mg1g7gflxMIoeSpXYSO5qwIL5ZqvoAD9H7J
CnQFO2xO3wrLq6TXH5Z7+0GWNghGk0GIOvF/ULHLWpsyhU5asKLK//MvORwQNHzL
b5LHG9uYeI+Jf12X4TI9qDaTCstiqM8vk1JPvgtSPJ9M62nRKY4ang==
-----END CERTIFICATE-----
EOF
```

重启 Docker 然后使用 docker login 登录：

```
[root@K8S-master01 ~]# docker login core.harbor.domain
Username: admin
Password:
Login Succeeded
```

如果报出证书不信任错误 x509: certificate signed by unknown authority，可以添加信任：

```
chmod 644 /etc/pki/ca-trust/extracted/pem/tls-ca-bundle.pem
```

将上述 ca.crt 添加到 /etc/pki/tls/certs/ca-bundle.crt 中：

```
chmod 444 /etc/pki/ca-trust/extracted/pem/tls-ca-bundle.pem
```

上传镜像。随便找一个镜像打上标记（tag），测试上传：

```
[root@K8S-master01 ~]# docker push core.harbor.domain/develop/busybox
The push refers to a repository [core.harbor.domain/develop/busybox]
8ac8bfaff55a: Pushed
latest: digest:
sha256:540f2e917216c5cfdf047b246d6b5883932f13d7b77227f09e03d42021e98941 size:
527
```

查看 Harbor Web，如图 3-26 所示。

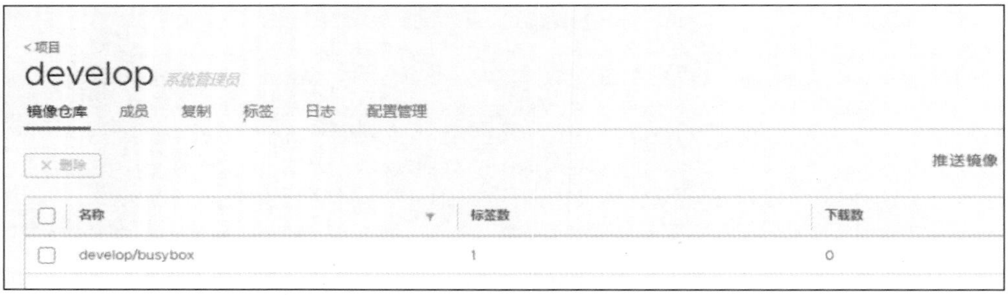

图 3-26 推送（Push）的镜像

3.8 安装 Prometheus+Grafana 到 K8S 集群中

本节使用的是 CoreOS 的 Prometheus-Operator 安装 Prometheus，本例监控包括监控 Etcd 集群，本例部署适用于二进制和 Kubeadm 安装方式，建议选择 Kubernetes 1.10 版本以上，其他版本自行测试。

Prometheus-Operator 项目地址：https://github.com/coreos/prometheus-operator/tree/master/contrib/kube-prometheus

3.8.1 修改配置信息

主要修改如下：

- Deploy 文件中 Etcd 证书文件位置，Kubeadm 安装方式无需修改。
- 修改 manifests/prometheus/prometheus-etcd.yaml 的 tlsConfig（kubeadm 安装方式无须修改）和 addresses（Etcd 地址）。
- 修改 alertmanager.yaml 文件的邮件报警配置和收件人配置（其他报警方式自行配置）。

3.8.2 一键安装 Prometheus

直接执行 Deploy 即可：

```
[root@K8S-master01 prometheus-operator]# ./deploy
namespace/monitoring created
secret/alertmanager-main created
secret/etcd-certs created
clusterrolebinding.rbac.authorization.K8S.io/prometheus-operator created
clusterrole.rbac.authorization.K8S.io/prometheus-operator created
serviceaccount/prometheus-operator created
service/prometheus-operator created
deployment.apps/prometheus-operator created
Waiting for Operator to register custom resource definitions...done!
clusterrolebinding.rbac.authorization.K8S.io/node-exporter created
clusterrole.rbac.authorization.K8S.io/node-exporter created
daemonset.extensions/node-exporter created
serviceaccount/node-exporter created
service/node-exporter created
clusterrolebinding.rbac.authorization.K8S.io/kube-state-metrics created
clusterrole.rbac.authorization.K8S.io/kube-state-metrics created
deployment.extensions/kube-state-metrics created
rolebinding.rbac.authorization.K8S.io/kube-state-metrics created
role.rbac.authorization.K8S.io/kube-state-metrics-resizer created
serviceaccount/kube-state-metrics created
service/kube-state-metrics created
secret/grafana-credentials created
secret/grafana-credentials unchanged
configmap/grafana-dashboard-definitions-0 created
configmap/grafana-dashboards created
configmap/grafana-datasources created
deployment.apps/grafana created
service/grafana created
service/etcd-K8S created
endpoints/etcd-K8S created
servicemonitor.monitoring.coreos.com/etcd-K8S created
configmap/prometheus-K8S-rules created
serviceaccount/prometheus-K8S created
servicemonitor.monitoring.coreos.com/alertmanager created
servicemonitor.monitoring.coreos.com/kube-apiserver created
servicemonitor.monitoring.coreos.com/kube-controller-manager created
servicemonitor.monitoring.coreos.com/kube-scheduler created
servicemonitor.monitoring.coreos.com/kube-state-metrics created
servicemonitor.monitoring.coreos.com/kubelet created
servicemonitor.monitoring.coreos.com/node-exporter created
servicemonitor.monitoring.coreos.com/prometheus-operator created
servicemonitor.monitoring.coreos.com/prometheus created
service/prometheus-K8S created
prometheus.monitoring.coreos.com/K8S created
role.rbac.authorization.K8S.io/prometheus-K8S created
role.rbac.authorization.K8S.io/prometheus-K8S created
role.rbac.authorization.K8S.io/prometheus-K8S created
clusterrole.rbac.authorization.K8S.io/prometheus-K8S created
rolebinding.rbac.authorization.K8S.io/prometheus-K8S created
rolebinding.rbac.authorization.K8S.io/prometheus-K8S created
rolebinding.rbac.authorization.K8S.io/prometheus-K8S created
clusterrolebinding.rbac.authorization.K8S.io/prometheus-K8S created
service/alertmanager-main created
alertmanager.monitoring.coreos.com/main created
```

> **注　意**
>
> 二进制安装方式注册时间可能会很长，Kubeadm 安装方式注册较迅速。

3.8.3 验证安装

查看 Pods：

```
[root@K8S-master01 prometheus-operator]# kubectl get po -n monitoring
NAME                                     READY   STATUS    RESTARTS   AGE
alertmanager-main-0                      2/2     Running   0          2m
alertmanager-main-1                      2/2     Running   0          1m
alertmanager-main-2                      2/2     Running   0          1m
grafana-59f56c4789-dzvgf                 1/1     Running   0          2m
kube-state-metrics-575464c49c-m8w4w      4/4     Running   0          2m
node-exporter-5kvxf                      2/2     Running   0          2m
node-exporter-66p7h                      2/2     Running   0          2m
node-exporter-clxzk                      2/2     Running   0          2m
node-exporter-hsgm8                      2/2     Running   0          2m
node-exporter-m5l24                      2/2     Running   0          2m
prometheus-K8S-0                         2/2     Running   0          2m
prometheus-K8S-1                         2/2     Running   0          2m
prometheus-operator-8597f9b976-2hvd5     1/1     Running   0          2m
```

查看 SVC：

```
[root@K8S-master01 prometheus-operator]# kubectl get svc -n !$
kubectl get svc -n monitoring
NAME                    TYPE        CLUSTER-IP       EXTERNAL-IP   PORT(S)                      AGE
alertmanager-main       NodePort    10.106.201.155   <none>        9093:30903/TCP               2m
alertmanager-operated   ClusterIP   None             <none>        9093/TCP,6783/TCP            2m
etcd-K8S                ClusterIP   None             <none>        2379/TCP                     2m
grafana                 NodePort    10.99.143.133    <none>        3000:30902/TCP               2m
kube-state-metrics      ClusterIP   None             <none>        8443/TCP,9443/TCP            2m
node-exporter           ClusterIP   None             <none>        9100/TCP                     2m
prometheus-K8S          NodePort    10.101.175.59    <none>        9090:30900/TCP               2m
prometheus-operated     ClusterIP   None             <none>        9090/TCP                     2m
prometheus-operator     ClusterIP   10.107.31.10     <none>        8080/TCP                     2m
```

一共开放了以下 3 个端口：

- alertmanager UI：30903
- grafana：30902
- prometheus UI：30900

3.8.4 访问测试

访问 Altermanager，查看 Status，如图 3-27 所示。

图 3-27　Altermanager Status 页面

访问 Prometheus，查看 Targets，如图 3-28 所示。

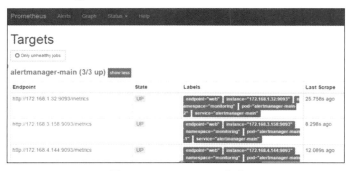

图 3-28　Prometheus 页面

访问 Grafana，查看资源使用，如图 3-29 所示。

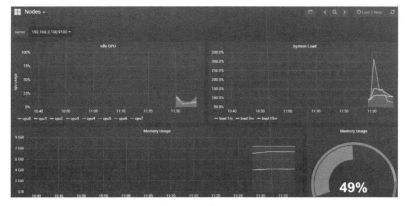

图 3-29　资源展示

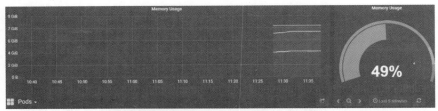

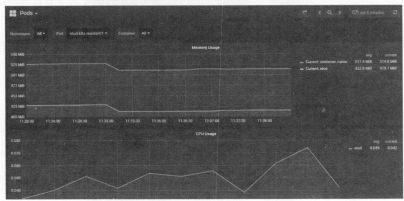

图 3-29（续）

登录邮箱，查看报警邮件，如图 3-30 所示。

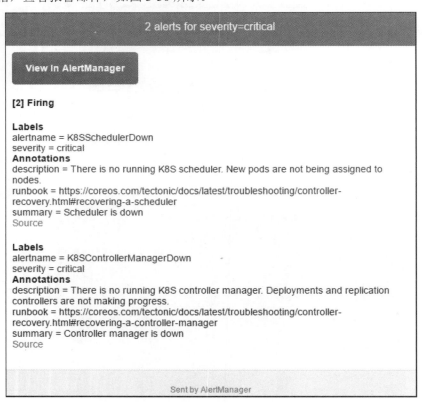

图 3-30　报警邮件

3.8.5 卸载

可以直接使用 teardown 脚本卸载，或者删除 monitoring 的命名空间：

```
[root@K8S-master01 prometheus-operator]# ./teardown
clusterrolebinding.rbac.authorization.K8S.io "node-exporter" deleted
clusterrole.rbac.authorization.K8S.io "node-exporter" deleted
daemonset.extensions "node-exporter" deleted
serviceaccount "node-exporter" deleted
service "node-exporter" deleted
clusterrolebinding.rbac.authorization.K8S.io "kube-state-metrics" deleted
clusterrole.rbac.authorization.K8S.io "kube-state-metrics" deleted
deployment.extensions "kube-state-metrics" deleted
rolebinding.rbac.authorization.K8S.io "kube-state-metrics" deleted
role.rbac.authorization.K8S.io "kube-state-metrics-resizer" deleted
serviceaccount "kube-state-metrics" deleted
service "kube-state-metrics" deleted
secret "grafana-credentials" deleted
configmap "grafana-dashboard-definitions-0" deleted
configmap "grafana-dashboards" deleted
configmap "grafana-datasources" deleted
deployment.apps "grafana" deleted
service "grafana" deleted
service "etcd-K8S" deleted
servicemonitor.monitoring.coreos.com "etcd-K8S" deleted
......
```

3.8.6 监控 ElasticSearch 集群

上一小节讲解了如何一键安装 Prometheus 监控，实际使用中仅使用资源监控和异常报警是远远不够的，还需要监控各类基础组件，如 ElasticSearch、RabbitMQ 等，并根据实际使用情况，对资源监控进行分类并设置报警策略，比如各类组件集群的状态、将不同的业务报警发送给不同的技术人员等。

本节将演示对 Kubernetes 集群外部的 ElasticSearch 集群监控，对于 Kubernetes 集群内部的 ElasticSearch 监控配置相同。

使用 Prometheus 进行业务监控，一般可分为以下步骤：

（1）创建 Endpoints 用于连接至外部服务。

（2）创建 Service 用于与 Endpoint 进行匹配。

（3）创建 Exporter 采集工具并配置上面创建的 Service 用于连接被监控业务。

（4）创建 Exporter 的 Service 用于被 Prometheus 发现并注册。

（5）创建 ServiceMonitor 将被监控资源注册至 Prometheus。

根据上述步骤，首先创建用于连接 Kubernetes 集群外部的 ElasticSearch 集群的 Endpoints 和 Service（集群内部的 ElasticSearch 一般都已经创建了 Service 和 Endpoints，可以直接使用，无须再次创建），修改 external-es.yaml 文件中的 Endpoints 的 addresses（地址）和 port（端口），然后创

建即可:

```
kind: Endpoints
apiVersion: v1
metadata:
  name: external-es
  namespace: logging
  labels:
    K8S-app: elasticsearch-logging
subsets:
  - addresses:
      - ip: 192.168.100.193
      - ip: 192.168.100.194
      - ip: 192.168.100.195
    ports:
      - port: 9200
        name: es
        protocol: TCP
---
apiVersion: v1
kind: Service
metadata:
  name: external-es
  namespace: logging
  labels:
    K8S-app: elasticsearch-logging
spec:
  type: ClusterIP
  clusterIP: None
  ports:
  - port: 9200
    protocol: TCP
    targetPort: 9200
    name: es
```

> **提 示**
>
> 当 Service 的名字和 Endpoint 的名字相同时，会自动建立联系。

一般情况下，在 Prometheus 监控中，对于数据的采集均使用一个被名为 Exporter 的采集工具，然后通过 Deployment 或者 DaemonSet 将此 Exporter 工具部署到集群中，用于数据的采集和汇总。比如采集宿主机节点信息的 exporter：node-exporter：

```
$ kubectl get po -n monitoring -l app=node-exporter
NAME                    READY   STATUS    RESTARTS   AGE
node-exporter-bpqx2     2/2     Running   0          51d
node-exporter-hrvjg     2/2     Running   2          51d
node-exporter-jlwsr     2/2     Running   0          42d
node-exporter-jns4f     2/2     Running   2          51d
node-exporter-lrnnw     2/2     Running   0          51d
node-exporter-pkttq     2/2     Running   0          51d
node-exporter-s6scl     2/2     Running   0          51d
node-exporter-s7lkm     2/2     Running   2          51d
node-exporter-sdt6b     2/2     Running   0          51d
```

以同样的方式创建 ElasticSearch 的 Exporter 用于采集 ElasticSearch 集群的各类监控数据。ElasticSearch Exporter 文件（es-exporter-deploy.yaml 文件）如下（需要注意的是，-es.uri=http://external-es.logging.svc:9200 参数中配置的外部 ElasticSearch 集群的 Service 地址，本例配置为上述创建的 Service）：

```yaml
apiVersion: extensions/v1beta1
kind: Deployment
metadata:
  labels:
    K8S-app: elasticsearch-exporter
    release: es-exporter
  name: es-exporter-elasticsearch-exporter
  namespace: monitoring
spec:
  progressDeadlineSeconds: 600
  replicas: 1
  revisionHistoryLimit: 10
  selector:
    matchLabels:
      K8S-app: elasticsearch-exporter
      release: es-exporter
  strategy:
    rollingUpdate:
      maxSurge: 1
      maxUnavailable: 0
    type: RollingUpdate
  template:
    metadata:
      creationTimestamp: null
      labels:
        K8S-app: elasticsearch-exporter
        release: es-exporter
    spec:
      containers:
      - command:
        - elasticsearch_exporter
        - -es.uri=http://external-es.logging.svc:9200
        - -es.all=true
        - -es.indices=true
        - -es.timeout=30s
        - -web.listen-address=:9108
        - -web.telemetry-path=/metrics
        image: justwatch/elasticsearch_exporter:1.0.2
        imagePullPolicy: IfNotPresent
        livenessProbe:
          failureThreshold: 3
          httpGet:
            path: /health
            port: http
            scheme: HTTP
          initialDelaySeconds: 30
          periodSeconds: 10
          successThreshold: 1
          timeoutSeconds: 10
```

```yaml
        name: elasticsearch-exporter
        ports:
        - containerPort: 9108
          name: http
          protocol: TCP
        readinessProbe:
          failureThreshold: 3
          httpGet:
            path: /health
            port: http
            scheme: HTTP
          initialDelaySeconds: 10
          periodSeconds: 10
          successThreshold: 1
          timeoutSeconds: 10
        resources: {}
        securityContext:
          capabilities:
            drop:
            - SETPCAP
            - MKNOD
            - AUDIT_WRITE
            - CHOWN
            - NET_RAW
            - DAC_OVERRIDE
            - FOWNER
            - FSETID
            - KILL
            - SETGID
            - SETUID
            - NET_BIND_SERVICE
            - SYS_CHROOT
            - SETFCAP
          procMount: Default
          readOnlyRootFilesystem: true
        terminationMessagePath: /dev/termination-log
        terminationMessagePolicy: File
      dnsPolicy: ClusterFirst
      restartPolicy: Always
      schedulerName: default-scheduler
      securityContext:
        runAsNonRoot: true
        runAsUser: 1000
      terminationGracePeriodSeconds: 30
```

上述步骤创建了用于采集数据的 Exporter，还需要创建一个 Exporter 的 Service，用于将运行 Exporter 的 Pod 暴露给 Prometheus，创建 Service 文件（es-exporter-svc.yaml）如下：

```yaml
apiVersion: v1
kind: Service
metadata:
  labels:
    K8S-app: elasticsearch-exporter
    release: es-exporter
  name: es-exporter-elasticsearch-exporter
```

```yaml
      namespace: monitoring
spec:
  ports:
  - name: http
    port: 9108
    protocol: TCP
    targetPort: 9108
  selector:
    K8S-app: elasticsearch-exporter
    release: es-exporter
  sessionAffinity: None
  type: ClusterIP
```

上述各类资源创建完成以后，需要创建 ServiceMonitor 将监控注册到 Prometheus 中，如果配置了自动发现则无须此步骤。创建 ServiceMonitor 文件（es-servicemonitor.yaml）如下：

```yaml
apiVersion: monitoring.coreos.com/v1
kind: ServiceMonitor
metadata:
  labels:
    K8S-app: elasticsearch-exporter
    release: es-exporter
  name: es-exporter-elasticsearch-exporter
  namespace: monitoring
spec:
  endpoints:
  - honorLabels: true
    interval: 10s
    path: /metrics
    port: http
    scheme: http
  jobLabel: es-exporter
  namespaceSelector:
    matchNames:
    - monitoring
  selector:
    matchLabels:
      K8S-app: elasticsearch-exporter
      release: es-exporter
```

当 ServiceMonitor 创建完成以后，对 ElasticSearch 集群的监控就完成了，此时可以通过访问 Prometheus 的 UI 界面查看 ElasticSearch 是否成功注册到了 Prometheus，访问上述部署的 Prometheus（上述部署的 Prometheus 的端口为 30900），查看 Target，如图 3-31 所示。

图 3-31　Prometheus UI 界面

在 Prometheus 的 UI 界面中能看到 es-exporter-elasticsearch-exporter 的 target 即表示监控成功。

添加完监控以后，需要在 Grafana 中添加监控 ElasticSearch 的 Dashboard，用于展示 ElasticSearch 数据。将登录之前创建的 Grafana 导入 Dashboard（ElasticSearch-Cluster-1555488325737.json），如图 3-32 所示。

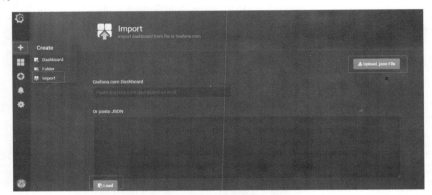

图 3-32　导入 Dashboard

对于 ElasticSearch 的监控包含了很多的监控项，比如：KPI、Shards、JVM 信息等，可以通过上述添加的 Dashboard 查看对应的监控数据，如图 3-33 所示。

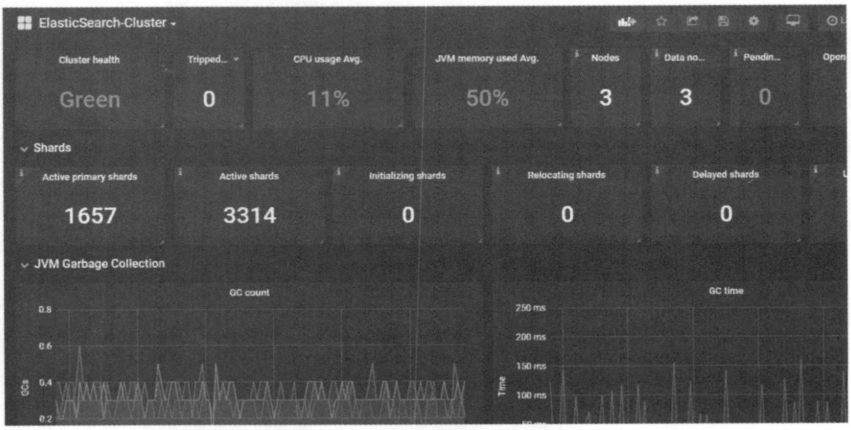

图 3-33　查看监控数据

通过上述步骤即完成了对 ElasticSearch 集群的监控，其他应用的监控，比如：RabbitMQ、Redis、Zookeeper 等配置均可参考上述步骤。有关采集数据的工具在开源社区 GitHub 上基本上都能找到对应的 Exporter，同样参考上述步骤即可完成对其他服务的监控。

3.8.7　监控报警配置实战

上一节讲解了 Prometheus 如何添加监控并通过 Grafana 展示图表，本节将演示对于一个新的监控如何添加报警策略，这也是监控体系中至关重要的一步。虽然之前一键安装 Prometheus 后也创建了一些基本的监控报警策略，但是在生产环境中还是要根据业务需求进行自定义监控报警。

通过 Prometheus 添加监控以后，可以在 Prometheus 的 Web 界面的 Graph 页面查看对应的业务数据，如查看上节监控的 ElasticSearch 的监控数据，如图 3-34 所示。

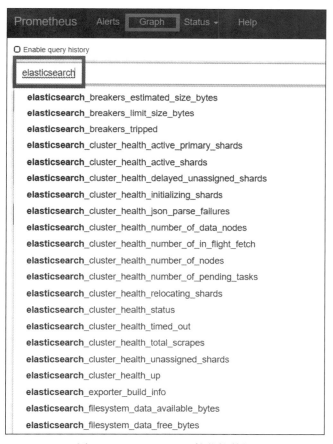

图 3-34　ElasticSearch 的监控数据

之后选择对应的监控指标，查看当前监控指标的数据，如查看当前 ElasticSearch 的集群状态，如图 3-35 所示。

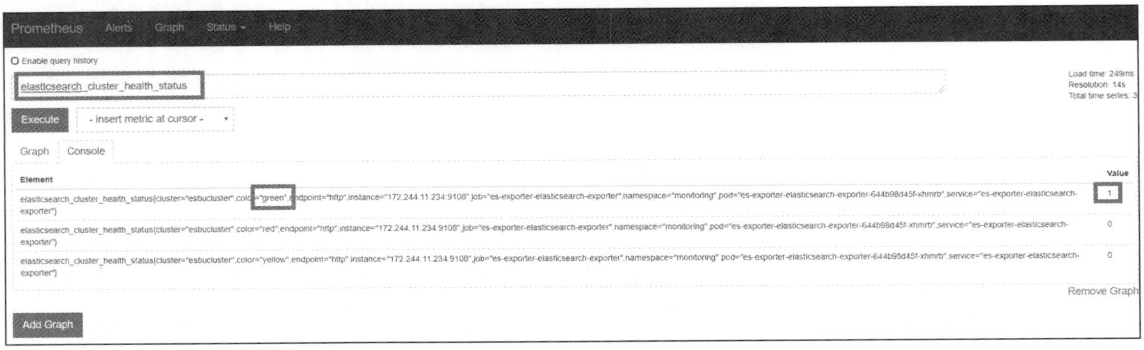

图 3-35 监控数据查看

可以看到，此时集群状态 green 为 1，red 和 yellow 为 0，说明此时集群状态为正常状态。通过这些监控指标可以获取到被监控业务的一些监控数据，然后通过这些数据，可以根据需求添加对应的报警策略，比如当集群状态的 yellow 值为 1 时触发报警，对应的报警策略如下：

```
- alert: Elastic_Cluster_Health_Yellow
  expr: elasticsearch_cluster_health_status{color="yellow"}==1
  for: 10m
  labels:
    severity: warning
    value: '{{$value}}'
  annotations:
    description: "Instance {{ $labels.instance }}: not all primary and replica shards are allocated in elasticsearch cluster {{ $labels.cluster }}."
    summary: "Instance {{ $labels.instance }}: not all primary and replica shards are allocated in elasticsearch cluster {{ $labels.cluster }}"
```

其中，需要注意的参数如下：

- alert　报警策略名称。
- expr　报警策略表达式。
- for　评估等待时间。
- annotations　报警信息，一般为报警内容。

上述为监控指标中的某一项指标的监控策略，可以根据上述方法添加其他监控项的报警策略，也可以使用已经定义好的监控策略文件，直接更新之前创建的报警策略：

```
$ kubectl replace -f prometheus-K8S-rules.yaml -n monitoring
```

该文件在末尾处添加了 ElasticSearch 集群关键的监控指标报警策略，当然读者也可以自行修改，然后更新之前创建的报警策略即可。更新报警策略配置以后，Prometheus 会自动更新该配置，此时可以在 Prometheus 的 Web 界面中看到对应的报警策略，如图 3-36 所示。

至此，完成了对 ElasticSearch 集群的监控和报警，对于其他应用的监控及报警配置基本相同，只需找到对应的 Exporter，然后根据 3.8.6 节的步骤添加监控即可。Prometheus 社区提供了丰富的 Exporter，涵盖了基础设施、中间件、网络设备等各个方面的监控功能。常见的 Exporter 如表 3-1 所示。

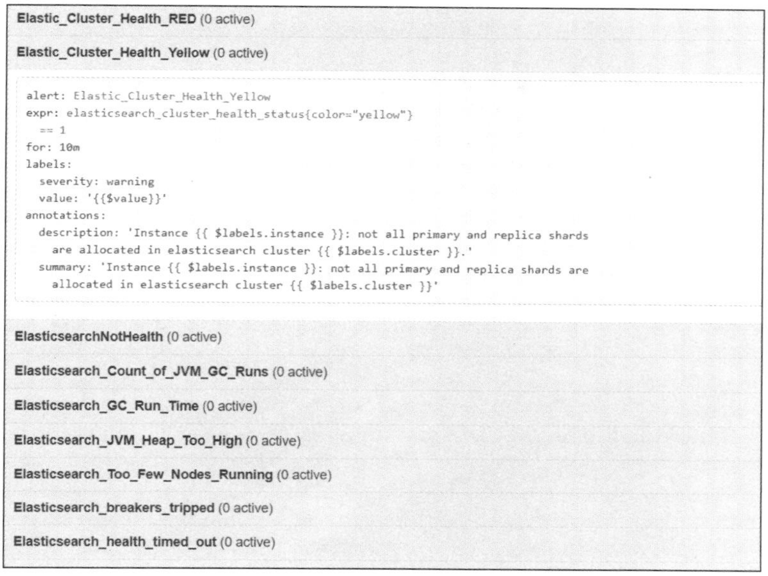

```
Elastic_Cluster_Health_RED (0 active)
Elastic_Cluster_Health_Yellow (0 active)

  alert: Elastic_Cluster_Health_Yellow
  expr: elasticsearch_cluster_health_status{color="yellow"}
    == 1
  for: 10m
  labels:
    severity: warning
    value: '{{$value}}'
  annotations:
    description: 'Instance {{ $labels.instance }}: not all primary and replica shards
      are allocated in elasticsearch cluster {{ $labels.cluster }}.'
    summary: 'Instance {{ $labels.instance }}: not all primary and replica shards are
      allocated in elasticsearch cluster {{ $labels.cluster }}'

ElasticsearchNotHealth (0 active)
Elasticsearch_Count_of_JVM_GC_Runs (0 active)
Elasticsearch_GC_Run_Time (0 active)
Elasticsearch_JVM_Heap_Too_High (0 active)
Elasticsearch_Too_Few_Nodes_Running (0 active)
Elasticsearch_breakers_tripped (0 active)
Elasticsearch_health_timed_out (0 active)
```

图 3-36　报警策略

表 3-1　常见的 Exporter

类型	Exporter
数据库	MySQL Exporter, Redis Exporter, MongoDB Exporter, MSSQL Exporter
硬件	Apcupsd Exporter，IoT Edison Exporter， IPMI Exporter, Node Exporter
消息队列	Beanstalkd Exporter, Kafka Exporter, NSQ Exporter, RabbitMQ Exporter
存储	Ceph Exporter, Gluster Exporter, HDFS Exporter, ScaleIO Exporter
HTTP 服务	Apache Exporter, HAProxy Exporter, Nginx Exporter
API 服务	AWS ECS Exporter， Docker Cloud Exporter, Docker Hub Exporter, GitHub Exporter
日志	Fluentd Exporter, Grok Exporter
监控系统	Collectd Exporter, Graphite Exporter, InfluxDB Exporter, Nagios Exporter, SNMP Exporter
其他	Blockbox Exporter, JIRA Exporter, Jenkins Exporter， Confluence Exporter

3.9　安装 EFK 到 K8S 集群中

EFK 是用于 Kubernetes 集群的日志收集工具，读者可以根据自己的业务情况，选择日志收集工具，比如还可以选择以 Sidecar 形式运行在 Pod 中的 Filebeat。本节介绍持久化安装 EFK 至 Kubernetes 集群，用于收集集群的相关日志，但是在生产环境中不建议将 EFK 安装到 Kubernetes 集群中，可以选择安装于集群之外，也可以将日志输出到公司外部的 ES 集群中。

3.9.1　对节点打标签（Label）

日志的采集使用 Fluentd，需要在每个收集日志的宿主机上部署一个 Fluentd 容器，然后对容器的日志进行收集，默认的日志输出在宿主机的/var/lib/docker 目录下。

为了收集 Fluentd 容器的日志，需要对收集日志的节点打标签，具体操作如下：

```
[root@K8S-master01 fluentd-elasticsearch]# kubectl label node
beta.kubernetes.io/fluentd-ds-ready=true --all
    node/K8S-master01 labeled
    node/K8S-master02 labeled
    node/K8S-master03 labeled
    node/K8S-node01 labeled
    node/K8S-node02 labeled
[root@K8S-master01 fluentd-elasticsearch]# kubectl get nodes --show-labels |
grep beta.kubernetes.io/fluentd-ds-ready
    K8S-master01    Ready     master    1d      v1.11.1
beta.kubernetes.io/arch=amd64,beta.kubernetes.io/fluentd-ds-ready=true,beta.ku
bernetes.io/os=linux,kubernetes.io/hostname=K8S-master01,node-role.kubernetes.
io/master=
    K8S-master02    Ready     master    1d      v1.11.1
beta.kubernetes.io/arch=amd64,beta.kubernetes.io/fluentd-ds-ready=true,beta.ku
bernetes.io/os=linux,kubernetes.io/hostname=K8S-master02,node-role.kubernetes.
io/master=
    K8S-master03    Ready     master    1d      v1.11.1
beta.kubernetes.io/arch=amd64,beta.kubernetes.io/fluentd-ds-ready=true,beta.ku
bernetes.io/os=linux,kubernetes.io/hostname=K8S-master03,node-role.kubernetes.
io/master=
    K8S-node01      Ready     <none>    7h      v1.11.1
beta.kubernetes.io/arch=amd64,beta.kubernetes.io/fluentd-ds-ready=true,beta.ku
bernetes.io/os=linux,kubernetes.io/hostname=K8S-node01
    K8S-node02      Ready     <none>    7h      v1.11.1
beta.kubernetes.io/arch=amd64,beta.kubernetes.io/fluentd-ds-ready=true,beta.ku
bernetes.io/os=linux,kubernetes.io/hostname=K8S-node02
```

3.9.2 创建持久化卷

本例安装使用 NFS 作为后端存储，可按需改为 GFS、CEPH。

在 NFS 上创建 ES 的存储目录（使用动态存储可以自动创建 PV，无须此步骤）：

```
[root@nfs es]# pwd
/nfs/es
[root@nfs es]# mkdir es{0..2}
```

三个 ES 实例创建三个目录，以此类推。

3.9.3 创建集群

修改 ingress.yaml 的 Kibana 域名后可直接进行创建：

```
[root@K8S-master01 efk-static]# kubectl apply -f .
namespace/logging configured
service/elasticsearch-logging created
serviceaccount/elasticsearch-logging created
clusterrole.rbac.authorization.K8S.io/elasticsearch-logging configured
```

```
    clusterrolebinding.rbac.authorization.K8S.io/elasticsearch-logging
configured
    statefulset.apps/elasticsearch-logging created
    persistentvolume/pv-es-0 configured
    persistentvolume/pv-es-1 configured
    persistentvolume/pv-es-2 configured
    configmap/fluentd-es-config-v0.1.1 created
    serviceaccount/fluentd-es created
    clusterrole.rbac.authorization.K8S.io/fluentd-es configured
    clusterrolebinding.rbac.authorization.K8S.io/fluentd-es configured
    daemonset.apps/fluentd-es-v2.0.2 created
    deployment.apps/kibana-logging created
    service/kibana-logging created
```

查看资源：

```
    [root@K8S-master01 efk-static]# kubectl get pv
    NAME          CAPACITY    ACCESS MODES    RECLAIM POLICY    STATUS      CLAIM
STORAGECLASS         REASON     AGE
    pv-es-0       4Gi         RWX             Recycle           Bound
logging/elasticsearch-logging-elasticsearch-logging-1    es-storage-class
59m
    pv-es-1       4Gi         RWX             Recycle           Available
es-storage-class              59m
    pv-es-2       4Gi         RWX             Recycle           Bound
logging/elasticsearch-logging-elasticsearch-logging-0    es-storage-class
59m
    [root@K8S-master01 efk-static]# kubectl get pvc -n logging
    NAME                                              STATUS   VOLUME    CAPACITY
ACCESS MODES   STORAGECLASS         AGE
    elasticsearch-logging-elasticsearch-logging-0     Bound    pv-es-2   4Gi
RWX            es-storage-class     1h
    elasticsearch-logging-elasticsearch-logging-1     Bound    pv-es-0   4Gi
RWX            es-storage-class     1h
```

```
    [root@K8S-master01 ~]# kubectl get pods -n logging
    NAME                                READY    STATUS     RESTARTS    AGE
    elasticsearch-logging-0             1/1      Running    2           3h
    elasticsearch-logging-1             1/1      Running    0           3h
    fluentd-es-v2.0.4-5jqhw             1/1      Running    0           10m
    fluentd-es-v2.0.4-fw5gk             1/1      Running    0           10m
    fluentd-es-v2.0.4-hm2tc             1/1      Running    0           10m
    fluentd-es-v2.0.4-nqqtm             1/1      Running    0           10m
    fluentd-es-v2.0.4-r5fgh             1/1      Running    0           10m
    kibana-logging-677854568-l6pvc      1/1      Running    0           3h
```

3.9.4　访问 Kibana

通过 Ingress 中配置的域名访问 Kibana，如图 3-37 所示。

按照图 3-37 配置 Kibana 后，单击 Create 按钮，等待 5 分钟左右再查看日志，即可看到 Pod 的输出日志，如图 3-38 所示。

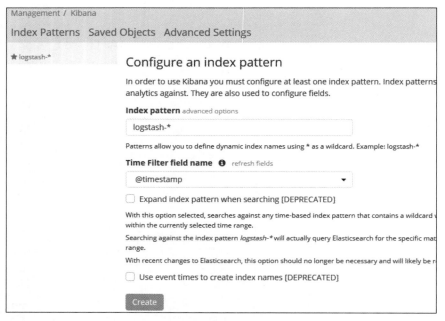

图 3-37　Kibana 页面

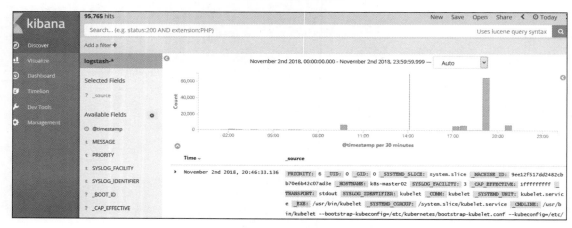

图 3-38　日志展示

3.10　小　结

本章讲解了企业中常用的中间件等工具的容器化安装，在实际使用时，对每个中间件等工具的容器化并非是必需的，可以根据自己的业务场景进行选择性容器化。比如笔者公司线下的测试环境，几乎全部都是采用容器化部署，具有非常好的可扩展性和迁移性，生产环境中，比如 Redis，一般情况下采用 Cluster 模式部署，很少出现同时宕机的情况。如果只是用来缓存会话，允许数据短暂性丢失，均可以在 Kubernetes 中部署。如果 Kubernetes 集群有高性能存储，几乎所有的应用都可以部署在 Kubernetes 集群中，实现统一管理。

第 4 章

持续集成与持续部署

本章主要介绍在生产环境中持续集成与持续部署的使用，主要通过实现 Jenkins 流水线脚本自动发布应用到 Kubernetes 集群当中。

4.1 CI/CD 介绍

CI（Continuous Integration，持续集成）/CD（Continuous Delivery，持续交付）是一种通过在应用开发阶段引入自动化来频繁向客户交付应用的方法。CI/CD 的核心概念是持续集成、持续交付和持续部署。作为一个面向开发和运营团队的解决方案，CI/CD 主要针对在集成新代码时所引发的问题（亦称"集成地狱"）。

具体而言，CI/CD 在整个应用生命周期内（从集成和测试阶段到交付和部署）引入了持续自动化和持续监控，这些关联的事务通常被称为"CI/CD 管道"，由开发和运维团队以敏捷方式协同支持。

4.1.1 CI 和 CD 的区别

CI/CD 中的 CI 指持续集成，它属于开发人员的自动化流程。成功的 CI 意味着应用代码的最新更改会定期构建、测试并合并到共享存储中。该解决方案可以解决在一次开发中有太多应用分支，从而导致相互冲突的问题。

CI/CD 中的 CD 指的是持续交付或持续部署，这些相关概念有时会交叉使用。两者都事关管道后续阶段的自动化，但它们有时也会单独使用，用于说明自动化程度。

持续交付通常是指开发人员对应用的更改会自动进行错误测试并上传到存储库（如 GitLab 或容器注册表），然后由运维团队将其部署到实时生产环境中，旨在解决开发和运维团队之间可见性及沟通较差的问题，因此持续交付的目的就是确保尽可能减少部署新代码时所需的工作量。

持续部署指的是自动将开发人员的更改从存储库发布到生产环境中以供客户使用，它主要为解决因手动流程降低应用交付速度，从而使运维团队超负荷的问题。持续部署以持续交付的优势为根基，实现了管道后续阶段的自动化。

CI/CD 既可能仅指持续集成和持续交付构成的关联环节，也可以指持续集成、持续交付和持续部署这三个方面构成的关联环节。更为复杂的是有时持续交付也包含了持续部署流程。

纠缠于这些语义其实并无必要，只需记得 CI/CD 实际上就是一个流程（通常表述为管道），用于在更大程度上实现应用开发的持续自动化和持续监控。

4.1.2 持续集成（CI）

现代应用开发的目标是让多位开发人员同时开发同一个应用的不同功能。但是，如果企业安排在一天内将所有分支源代码合并在一起，最终可能导致工作繁琐、耗时，而且需要手动完成。这是因为当一位独立工作的开发人员对应用进行更改时，有可能会有其他开发人员同时进行更改，从而引发冲突。

持续集成可以帮助开发人员更加频繁地将代码更改合并到共享分支或主干中。一旦开发人员对应用所做的更改被合并，系统就会通过自动构建应用并运行不同级别的自动化测试（通常是单元测试和集成测试）来验证这些更改，确保更改没有对应用造成破坏。这意味着测试内容涵盖了从类和函数到构成整个应用的不同模块，如果自动化测试发现新代码和现有代码之间有冲突，持续集成（CI）可以更加轻松地快速修复这些错误。

4.1.3 持续交付（CD）

完成持续集成中构建单元测试和集成测试的自动化流程后，通过持续交付可自动将已验证的代码发布到存储库。为了实现高效的持续交付流程，务必要确保持续交付已内置于开发管道。持续交付的目标是拥有一个可随时部署到生产环境的代码库。

在持续交付中，每个阶段（从代码更改的合并到生产就绪型构建版本的交付）都涉及测试自动化和代码发布自动化。在流程结束时，运维团队可以快速、轻松地将应用部署到生产环境中。

4.1.4 持续部署

对于一个成熟的 CI/CD 管道来说，最后的阶段是持续部署。作为持续交付（自动将生产就绪型构建版本发布到代码存储库）的延伸，持续部署可以自动将应用发布到生产环境中。由于生产之前的管道阶段没有手动门控，因此持续部署在很大程度上都得依赖于精心设计的测试自动化。

实际上，持续部署意味着开发人员对应用的更改在编写后的几分钟内就能生效，这更加便于持续接收和整合用户反馈。总而言之，所有这些 CI/CD 的关联步骤都有助于降低应用的部署风险，因此更便于以小件的方式（非一次性）发布对应用的更改。不过，由于还需要编写自动化测试以适应 CI/CD 管道中的各种测试和发布阶段，因此前期投资会很大。

4.2 Jenkins 流水线介绍

本节主要讲解 Jenkins 的新功能——Jenkins 流水线（Pipeline）的使用，首先介绍流水线的概念和类型，然后讲解流水线的基本语法和一些例子。

4.2.1 什么是流水线

Jenkins 流水线（或 Pipeline）是一套插件，它支持实现并把持续提交流水线（Continuous Delivery Pipeline）集成到 Jenkins。

持续提交流水线（Continuous Delivery Pipeline）会经历一个复杂的过程：从版本控制、向用户和客户提交软件，软件的每次变更（提交代码到仓库）到软件发布（Release）。这个过程包括以一种可靠并可重复的方式构建软件，以及通过多个测试和部署阶段来开发构建好的软件（称为 Build）。

流水线提供了一组可扩展的工具，通过流水线语法对从简单到复杂的交付流水线作为代码进行建模，Jenkins 流水线的定义被写在一个文本文件中，一般为 Jenkinsfile，该文件"编制"了整个构建软件的过程，该文件一般也可以被提交到项目的代码仓库中，在 Jenkins 中可以直接引用。这是流水线即代码的基础，将持续提交流水线作为应用程序的一部分，像其他代码一样进行版本化和审查。创建 Jenkinsfile 并提交到代码仓库中的好处如下：

- 自动地为所有分支创建流水线构建过程。
- 在流水线上进行代码复查/迭代。
- 对流水线进行审计跟踪。
- 流水线的代码可以被项目的多个成员查看和编辑

4.2.2 Jenkins 流水线概念

流水线主要分为以下几种区块：Pipeline、Node、Stage、Step 等。

1. Pipeline（流水线）

Pipeline 是用户定义的一个持续提交（CD）流水线模型。流水线的代码定义了整个的构建过程，包括构建、测试和交付应用程序的阶段。另外，Pipeline 块是声明式流水线语法的关键部分。

2. Node（节点）

Node（节点）是一个机器，它是 Jenkins 环境的一部分，另外，Node 块是脚本化流水线语法的关键部分。

3. Stage（阶段）

Stage 块定义了在整个流水线的执行任务中概念不同的子集（比如 Build、Test、Deploy 阶段），它被许多插件用于可视化 Jenkins 流水线当前的状态/进展。

4. Step（步骤）

本质上是指通过一个单一的任务告诉 Jenkins 在特定的时间点需要做什么，比如要执行 shell 命令，可以使用 sh SHELL_COMMAND。

4.2.3 声明式流水线

在声明式流水线语法中，Pipeline 块定义了整个流水线中完成的所有工作，比如：

```
Jenkinsfile (Declarative Pipeline)
pipeline {
    agent any
    stages {
        stage('Build') {
            steps {
                //
            }
        }
        stage('Test') {
            steps {
                //
            }
        }
        stage('Deploy') {
            steps {
                //
            }
        }
    }
}
```

说明

- agentany：在任何可用的代理上执行流水线或它的任何阶段。
- stage('Build')：定义 Build 阶段。
- steps：执行某阶段相关的步骤。

4.2.4 脚本化流水线

在脚本化流水线语法中，会有一个或多个 Node（节点）块在整个流水线中执行核心工作，比如：

```
Jenkinsfile (Scripted Pipeline)
node {
    stage('Build') {
        //
    }
    stage('Test') {
        //
```

```
    }
    stage('Deploy') {
        //
    }
}
```

说明

- node：在任何可用的代理上执行流水线或它的任何阶段。
- stage('Build')：定义 build 阶段。stage 块在脚本化流水线语法中是可选的，然而在脚本化流水线中实现 stage 块，可以清楚地在 Jenkins UI 界面中显示每个 stage 的任务子集。

4.2.5 流水线示例

一个以声明式流水线的语法编写的 Jenkinsfile 文件如下：

```
Jenkinsfile (Declarative Pipeline)
pipeline {
    agent any
    stages {
        stage('Build') {
            steps {
                sh 'make'
            }
        }
        stage('Test'){
            steps {
                sh 'make check'
                junit 'reports/**/*.xml'
            }
        }
        stage('Deploy') {
            steps {
                sh 'make publish'
            }
        }
    }
}
```

常用参数说明

- pipeline 是声明式流水线的一种特定语法，定义了包含执行整个流水线的所有内容和指令。
- agent 是声明式流水线的一种特定语法，指示 Jenkins 为整个流水线分配一个执行器（在节点上）和工作区。
- stage 是一个描述流水线阶段的语法块，在脚本化流水线语法中，stage（阶段）块是可选的。
- steps 是声明式流水线的一种特定语法，它描述了在这个 stage 中要运行的步骤。
- sh 是一个执行给定 shell 命令的流水线 step（步骤）。
- junit 是一个聚合测试报告的流水线 step（步骤）。

- node 是脚本化流水线的一种特定语法，它指示 Jenkins 在任何可用的代理/节点上执行流水线，这实际等同于声明式流水线特定语法的 agent。注：后续的例子中要用到。

上述声明式流水线等同于以下脚本式流水线：

```
Jenkinsfile (Scripted Pipeline)
node {
    stage('Build') {
        sh 'make'
    }
    stage('Test') {
        sh 'make check'
        junit 'reports/**/*.xml'
    }
    stage('Deploy') {
        sh 'make publish'
    }
}
```

4.3 Pipeline 语法

本节主要从流水线的两种类型出发讲解 Pipeline 的语法。

4.3.1 声明式流水线

声明式流水线是在流水线子系统之上提供了一种更简单、更有主见的语法。

所有有效的声明式流水线必须包含在一个 Pipeline 块中，比如以下是一个 Pipeline 块的格式：

```
pipeline {
    /* insert Declarative Pipeline here */
}
```

在声明式流水线中有效的基本语句和表达式遵循与 Groovy 的语法同样的规则，但有以下例外：

- 流水线顶层必须是一个 block，pipeline{}。
- 没有分号作为语句分隔符，每条语句都在自己的行上。
- 块只能由 Sections、Directives、Steps 或 assignment statements 组成。

1. Sections

声明式流水线中的 Sections 通常包含一个或多个 agent、Stages、post、Directives 和 Steps，本节首先介绍 agent、Stages、post，有关 Directives 和 Steps 的说明见下一小节。

（1）agent

agent 部分指定了整个流水线或特定的部分，在 Jenkins 环境中执行的位置取决于 agent 区域的位置，该部分必须在 pipeline 块的顶层被定义，但是 stage 级别的使用是可选的。

①参数

为了支持可能有的各种各样的流水线，agent 部分支持一些不同类型的参数，这些参数应用在 pipeline 块的顶层，或 Stage 指令内部。

- any：在任何可用的代理上执行流水线或 stage。例如：agent any。
- none：当在 pipeline 块的顶部没有全局 agent，该参数将会被分配到整个流水线的运行中，并且每个 stage 部分都需要包含它自己的 agent，比如：agent none。
  在提供了标签的 Jenkins 环境中可用代理上执行流水线或 stage。例如：agent { label 'my-defined-label'}。
- node：agent { node { label 'labelName'} } 和 agent { label 'labelName' }一样，但是 node 允许额外的选项（比如 customWorkspace）。
- dockerfile：执行流水线或 stage，使用从源码包含的 Dockerfile 所构建的容器。为了使用该选项，Jenkinsfile 必须从多个分支流水线中加载，或者加载 Pipelinefrom SCM（下面章节会涉及）。通常，这是源码根目录下的 Dockerfile:agent { dockerfile true }。如果在其他目录下构建 Dockerfile，使用 dir 选择：agent { dockerfile { dir 'someSubDir'} }。如果 Dockerfile 有另一个名字，可以使用 filename 选项指定该文件名。也可以传递额外的参数到 dockerbuild，使用 additionalBuildArgs 选项提交，比如：agent { dockerfile { additionalBuildArgs '--build-arg foo=bar' } }。例如一个带有 build/Dockerfile.build 的仓库，在构建时期望一个参数 version：

```
agent {
    dockerfile {
        filename 'Dockerfile.build'
        dir 'build'
        label 'my-defined-label'
        additionalBuildArgs '--build-arg version=1.0.2'
    }
}
```

- docker：使用给定的容器执行流水线或 stage。该容器将在预置的 node（节点）上，或在由 label 参数指定的节点上，动态地接受基于 Docker 的流水线。Docker 也可以接受 args 参数，该参数可能包含直接传递到 dockerrun 调用的参数及 alwaysPull 选项，alwaysPul 选项强制 dockerpull，即使镜像（image）已经存在。比如：agent { docker 'maven:3-alpine' }或：

```
agent{
    docker{
        image 'maven:3-alpine'
        label 'my-defined-label'
        args '-v /tmp:/tmp'
    }
}
```

②常见选项

- label：一个字符串，该标签用于运行流水线或个别 stage。该选项对 node、docker 和 dockerfile 可用，node 必须选择该选项。
- customWorkspace：一个字符串，在自定义工作区运行流水线或 stage。它可以是相对路径，也可以是绝对路径，该选项对 node、docker 和 dockerfile 可用。比如：

```
agent {
    node {
        label 'my-defined-label'
        customWorkspace '/some/other/path'
    }
}
```

- reuseNode：一个布尔值，默认为 false。如果是 true，则在流水线顶层指定的节点上运行该容器。这个选项对 docker 和 dockerfile 有用，并且只有当使用在个别 stage 的 agent 上才会有效。

③示例

示例 1：在 maven:3-alpine（agent 中定义）的新建容器上执行定义在流水线中的所有步骤。

```
Jenkinsfile (Declarative Pipeline)
pipeline {
    agent { docker 'maven:3-alpine' }
    stages {
        stage('Example Build') {
            steps {
                sh 'mvn -B clean verify'
            }
        }
    }
}
```

示例 2：本示例在流水线顶层定义 agentnone，确保<<../glossary#executor, an Executor>> 没有被分配。使用 agentnode 也会强制 stage 部分包含它自己的 agent 部分。在 stage('Example Build')部分使用 maven:3-alpine 执行该阶段步骤，在 stage('Example Test')部分使用 openjdk:8-jre 执行该阶段步骤。

```
Jenkinsfile (Declarative Pipeline)
pipeline {
    agent none
    stages {
        stage('Example Build') {
            agent { docker 'maven:3-alpine' }
            steps {
                echo 'Hello, Maven'
                sh 'mvn --version'
            }
        }
        stage('Example Test') {
            agent { docker 'openjdk:8-jre' }
            steps {
                echo 'Hello, JDK'
                sh 'java -version'
            }
        }
    }
}
```

（2）post

post 部分定义一个或多个 steps，这些阶段根据流水线或 stage 的完成情况而运行（取决于流水线中 post 部分的位置）。post 支持以下 post-condition 块之一：

- always：无论流水线或 stage 的完成状态如何，都允许在 post 部分运行该步骤。
- changed：只有当前流水线或 stage 的完成状态与它之前的运行不同时，才允许在 post 部分运行该步骤。
- failure：只有当前流水线或 stage 的完成状态为失败（failure），才允许在 post 部分运行该步骤，通常这时在 Web 界面中显示为红色。
- success：当前状态为成功（success），执行 post 步骤，通常在 Web 界面中显示为蓝色或绿色。
- unstable：当前状态为不稳定（unstable），执行 post 步骤，通常由于测试失败或代码违规等造成，在 Web 界面中显示为黄色。
- aborted：当前状态为放弃（aborted），执行 post 步骤，通常由于流水线被手动放弃触发，这时在 Web 界面中显示为灰色。

示例：一般情况下 post 部分放在流水线的底部，比如本实例，无论 stage 的完成状态如何，都会输出一条 I will always say Hello again!信息。

```
Jenkinsfile (Declarative Pipeline)
pipeline {
    agent any
    stages {
        stage('Example') {
            steps {
                echo 'Hello World'
            }
        }
    }
    post {
        always {
            echo 'I will always say Hello again!'
        }
    }
}
```

（3）stages

stages 包含一个或多个 stage 指令，stages 部分是流水线描述的大部分工作（work）的位置。建议 stages 至少包含一个 stage 指令，用于持续交付过程的某个离散的部分，比如构建、测试或部署。

示例：本示例的 stages 包含一个名为 Example 的 stage，该 stage 执行 echo 'Hello World'命令输出 Hello World 信息。

```
Jenkinsfile (Declarative Pipeline)
pipeline {
    agent any
    stages {
        stage('Example') {
            steps {
                echo 'Hello World'
```

```
        }
      }
    }
}
```

（4）steps

steps 部分在给定的 stage 指令中执行的一个或多个步骤。

示例：在 steps 定义执行一条 shell 命令。

```
Jenkinsfile (Declarative Pipeline)
pipeline {
    agent any
    stages {
        stage('Example') {
            steps {
                echo 'Hello World'
            }
        }
    }
}
```

2. Directives

Directives 用于一些执行 stage 时的条件判断，主要分为 environment、options、parameters、triggers、stage、tools、input、when 等，这里仅对常用的 environment、parameters、stage 和 when 进行介绍。

（1）environment

environment 制定一个键-值对（key-value pair）序列，该序列将被定义为所有步骤的环境变量，或者是特定 stage 的步骤，这取决于 environment 指令在流水线内的位置。

该指令支持一个特殊的方法 credentials()，该方法可用于在 Jenkins 环境中通过标识符访问预定义的凭证。对于类型为 Secret Text 的凭证，credentials()将确保指定的环境变量包含秘密文本内容，对于类型为 Standardusernameandpassword 的凭证，指定的环境变量为 username:password，并且两个额外的环境变量将被自动定义，分别为 MYVARNAME_USR 和 MYVARNAME_PSW。

示例：

```
Jenkinsfile (Declarative Pipeline)
pipeline {
    agent any
    environment {
        CC = 'clang'
    }
    stages {
        stage('Example') {
            environment {
                AN_ACCESS_KEY = credentials('my-prefined-secret-text')
            }
            steps {
                sh 'printenv'
            }
        }
    }
```

}
```

上述示例的顶层流水线块中使用的 environment 指令将适用于流水线中的所有步骤。在 stage 中定义的 environment 指令只会适用于 stage 中的步骤。其中 stage 中的 environment 使用的是 credentials 预定义的凭证。

（2）parameters

parameters 提供了一个用户在触发流水线时应该提供的参数列表，这些用户指定参数的值可以通过 params 对象提供给流水线的 step（步骤）。

**可用参数**

- string：字符串类型的参数，例如：parameters { string(name: 'DEPLOY_ENV', defaultValue: 'staging', description: '') }。
- booleanParam：布尔参数，例如：parameters { booleanParam(name: 'DEBUG_BUILD', defaultValue: true, description: '') }。

示例：定义 string 类型的变量，并在 steps 中引用。

```
Jenkinsfile (Declarative Pipeline)
pipeline {
    agent any
    parameters {
        string(name: 'PERSON', defaultValue: 'Mr Jenkins', description: 'Who should I say hello to?')
    }
    stages {
        stage('Example') {
            steps {
                echo "Hello ${params.PERSON}"
            }
        }
    }
}
```

（3）stage

stage 指定在 stages 部分流水线所做的工作都将封装在一个或多个 stage 指令中。

示例：

```
Jenkinsfile (Declarative Pipeline)
pipeline {
    agent any
    stages {
        stage('Example') {
            steps {
                echo 'Hello World'
            }
        }
    }
}
```

（4）when

when 指令允许流水线根据给定的条件决定是否应该执行 stage。when 指令必须包含至少一个
```

条件。如果 when 包含多个条件，所有的子条件必须返回 True，stage 才能执行。

①内置条件

- branch：当正在构建的分支与给定的分支匹配时，执行这个 stage，例如：when { branch 'master' }。注意，branch 只适用于多分支流水线。
- environment：当指定的环境变量和给定的变量匹配时，执行这个 stage，例如：when { environment name: 'DEPLOY_TO', value: 'production' }。
- expression：当指定的 Groovy 表达式评估为 True，执行这个 stage，例如：when { expression { return params.DEBUG_BUILD } }。
- not：当嵌套条件出现错误时，执行这个 stage，必须包含一个条件，例如：when { not { branch 'master' } }。
- allOf：当所有的嵌套条件都正确时执行这个 stage，必须包含至少一个条件，例如：when { allOf { branch 'master'; environment name: 'DEPLOY_TO', value: 'production' } }
- anyOf：当至少有一个嵌套条件为 True 时执行这个 stage，例如：when { anyOf { branch 'master'; branch 'staging' } }。

②在进入 stage 的 agent 前评估 when

默认情况下，如果定义了某个 stage 的 agent，在进入该 stage 的 agent 后，该 stage 的 when 条件才会被评估。但是可以通过在 when 块中指定 beforeAgent 选项来更改此选项。如果 beforeAgent 被设置为 True，那么就会首先对 when 条件进行评估，并且只有在 when 条件验证为真时才会进入 agent。

③示例

示例 1：当 branch 为 production 时才会执行名为 Example Deploy 的 stage。

```
Jenkinsfile (Declarative Pipeline)
pipeline {
    agent any
    stages {
        stage('Example Build') {
            steps {
                echo 'Hello World'
            }
        }
        stage('Example Deploy') {
            when {
                branch 'production'
            }
            steps {
                echo 'Deploying'
            }
        }
    }
}
```

示例 2：当 branch 为 production，environment 的 DEPLOY_TO 为 production 才会执行名为 Example Deploy 的 stage。

```
Jenkinsfile (Declarative Pipeline)
pipeline {
    agent any
    stages {
        stage('Example Build') {
            steps {
                echo 'Hello World'
            }
        }
        stage('Example Deploy') {
            when {
                branch 'production'
                environment name: 'DEPLOY_TO', value: 'production'
            }
            steps {
                echo 'Deploying'
            }
        }
    }
}
```

示例 3：当 branch 为 production 并且 DEPLOY_TO 为 production 时才会执行名为 Example Deploy 的 stage。

```
Jenkinsfile (Declarative Pipeline)
pipeline {
    agent any
    stages {
        stage('Example Build') {
            steps {
                echo 'Hello World'
            }
        }
        stage('Example Deploy') {
            when {
                allOf {
                    branch 'production'
                    environment name: 'DEPLOY_TO', value: 'production'
                }
            }
            steps {
                echo 'Deploying'
            }
        }
    }
}
```

示例 4：当 DEPLOY_TO 等于 production 或者 staging 时才会执行名为 Example Deploy 的 stage。

```
Jenkinsfile (Declarative Pipeline)
pipeline {
    agent any
    stages {
        stage('Example Build') {
            steps {
```

```
                echo 'Hello World'
            }
        }
        stage('Example Deploy') {
            when {
                branch 'production'
                anyOf {
                    environment name: 'DEPLOY_TO', value: 'production'
                    environment name: 'DEPLOY_TO', value: 'staging'
                }
            }
            steps {
                echo 'Deploying'
            }
        }
    }
}
```

示例 5：当 BRANCH_NAME 为 production 或者 staging，并且 DEPLOY_TO 为 production 或者 staging 时才会执行名为 Example Deploy 的 stage。

```
Jenkinsfile (Declarative Pipeline)
pipeline {
    agent any
    stages {
        stage('Example Build') {
            steps {
                echo 'Hello World'
            }
        }
        stage('Example Deploy') {
            when {
                expression { BRANCH_NAME ==~ /(production|staging)/ }
                anyOf {
                    environment name: 'DEPLOY_TO', value: 'production'
                    environment name: 'DEPLOY_TO', value: 'staging'
                }
            }
            steps {
                echo 'Deploying'
            }
        }
    }
}
```

示例 6：在进行 agent 前执行判断，当 branch 为 production 时才会进行该 agent。

```
Jenkinsfile (Declarative Pipeline)
pipeline {
    agent none
    stages {
        stage('Example Build') {
            steps {
                echo 'Hello World'
            }
```

```
    }
    stage('Example Deploy') {
        agent {
            label "some-label"
        }
        when {
            beforeAgent true
            branch 'production'
        }
        steps {
            echo 'Deploying'
        }
    }
}
```

3. steps

Steps 包含一个完整的 script 步骤列表。Script 步骤需要 scripted-pipeline 块并在声明式流水线中执行。对于大多数用例来说，script 步骤并不是必要的。

示例：在 steps 添加 script 进行 for 循环。

```
Jenkinsfile (Declarative Pipeline)
pipeline {
    agent any
    stages {
        stage('Example') {
            steps {
                echo 'Hello World'

                script {
                    def browsers = ['chrome', 'firefox']
                    for (int i = 0; i < browsers.size(); ++i) {
                        echo "Testing the ${browsers[i]} browser"
                    }
                }
            }
        }
    }
}
```

4.3.2 脚本化流水线

脚本化流水线与声明式流水线一样都是建立在底层流水线的子系统上，与声明式流水线不同的是，脚本化流水线实际上是由 Groovy 构建。Groovy 语言提供的大部分功能都可以用于脚本化流水线的用户，这意味着它是一个非常有表现力和灵活的工具，可以通过它编写持续交付流水线。

脚本化流水线和其他传统脚本一致都是从 Jenkinsfile 的顶部开始向下串行执行，因此其提供的流控制也取决于 Groovy 表达式，比如：if/else 条件：

```
Jenkinsfile (Scripted Pipeline)
node {
    stage('Example') {
```

```
            if (env.BRANCH_NAME == 'master') {
                echo 'I only execute on the master branch'
            } else {
                echo 'I execute elsewhere'
            }
        }
    }
```

另一种方法是使用 Groovy 的异常处理支持来管理脚本化流水线的流控制,无论遇到什么原因的失败,它们都会抛出一个异常,处理错误的行为必须使用 Groovy 中的 try/catch/finally 块,例如:

```
Jenkinsfile (Scripted Pipeline)
node {
    stage('Example') {
        try {
            sh 'exit 1'
        }
        catch (exc) {
            echo 'Something failed, I should sound the klaxons!'
            throw
        }
    }
}
```

4.4 Jenkinsfile 的使用

上面讲过流水线支持两种语法,即声明式和脚本式,这两种语法都支持构建持续交付流水线。并且都可以用来在 Web UI 或 Jenkinsfile 中定义流水线,不过通常将 Jenkinsfile 放置于代码仓库中。

创建一个 Jenkinsfile 并将其放置于代码仓库中,有以下好处:

- 方便对流水线上的代码进行复查/迭代。
- 对管道进行审计跟踪。
- 流水线真正的源代码能够被项目的多个成员查看和编辑。

本节主要介绍 Jenkinsfile 常见的模式以及演示 Jenkinsfile 的一些特例。

4.4.1 创建 Jenkinsfile

Jenkinsfile 是一个文本文件,它包含了 Jenkins 流水线的定义并被用于源代码控制。以下流水线实现了 3 个基本的持续交付:

```
Jenkinsfile (Declarative Pipeline)
pipeline {
    agent any

    stages {
        stage('Build') {
            steps {
```

```
            echo 'Building..'
        }
    }
    stage('Test') {
        steps {
            echo 'Testing..'
        }
    }
    stage('Deploy') {
        steps {
            echo 'Deploying....'
        }
    }
}
```

对应的脚本式流水线如下：

```
Jenkinsfile (Scripted Pipeline)
node {
    stage('Build') {
        echo 'Building....'
    }
    stage('Test') {
        echo 'Testing....'
    }
    stage('Deploy') {
        echo 'Deploying....'
    }
}
```

> **注　意**
>
> 不是所有的流水线都有相同的三个阶段。

1. 构建

对于许多项目来说，流水线中工作（work）的开始就是构建（build），这个阶段的主要工作是进行源代码的组装、编译或打包。Jenkinsfile 文件不是替代现有的构建工具，如 GNU/Make、Maven、Gradle 等，可以视其为一个将项目开发周期的多个阶段（构建、测试、部署等）绑定在一起的粘合层。

Jenkins 有许多插件用于构建工具，假设系统为 Unix/Linux，只需要从 shell 步骤（sh）调用 make 即可进行构建，Windows 系统可以使用 bat：

```
Jenkinsfile (Declarative Pipeline)
pipeline {
    agent any

    stages {
        stage('Build') {
            steps {
                sh 'make'
                archiveArtifacts artifacts: '**/target/*.jar', fingerprint: true
```

```
            }
        }
    }
}
```

说明

- steps 的 shmake 表示如果命令的状态码为 0，则继续，为非零则失败。
- archiveArtifacts 用于捕获构建后生成的文件。

对应的脚本式流水线如下：

```
Jenkinsfile (Scripted Pipeline)
node {
    stage('Build') {
        sh 'make'
        archiveArtifacts artifacts: '**/target/*.jar', fingerprint: true
    }
}
```

2. 测试

运行自动化测试是任何成功的持续交付过程中的重要组成部分，因此 Jenkins 有许多测试记录、报告和可视化工具，这些工具都是由插件提供。下面的例子将使用 JUnit 插件提供的 junit 工具进行测试。

在下面的例子中，如果测试失败，流水线就会被标记为不稳定，这时 Web 界面中的球就显示为黄色。基于记录的测试报告，Jenkins 也可以提供历史趋势分析和可视化：

```
Jenkinsfile (Declarative Pipeline)
pipeline {
    agent any

    stages {
        stage('Test') {
            steps {
                /* `make check` returns non-zero on test failures,
                 * using `true` to allow the Pipeline to continue nonetheless
                 */
                sh 'make check || true'
                junit '**/target/*.xml'
            }
        }
    }
}
```

说明

- 当 sh 步骤状态码为 0 时，调用 junit 进行测试。
- junit 捕获并关联匹配 **/target/*.xml 的 Junit XML 文件。

对应的脚本式流水线如下：

```
Jenkinsfile (Scripted Pipeline)
node {
```

```
/* .. snip .. */
stage('Test') {
    /* `make check` returns non-zero on test failures,
     * using `true` to allow the Pipeline to continue nonetheless
     */
    sh 'make check || true'
    junit '**/target/*.xml'
}
/* .. snip .. */
```

3. 部署

当编译构建和测试都通过后，就会将编译生成的包推送到生产环境中，从本质上讲，Deploy 阶段只可能发生在之前的阶段都成功完成后才会进行，否则流水线会提前退出：

```
Jenkinsfile (Declarative Pipeline)
pipeline {
    agent any

    stages {
        stage('Deploy') {
            when {
              expression {
                currentBuild.result == null || currentBuild.result == 'SUCCESS'
              }
            }
            steps {
                sh 'make publish'
            }
        }
    }
}
```

> **说 明**
>
> 当前 build 结果为 SUCCESS 时，执行 publish。

对应的脚本式流水线如下：

```
Jenkinsfile (Scripted Pipeline)
node {
    /* .. snip .. */
    stage('Deploy') {
        if (currentBuild.result == null || currentBuild.result == 'SUCCESS') {
            sh 'make publish'
        }
    }
    /* .. snip .. */
}
```

4.4.2 处理 Jenkinsfile

本节主要介绍 Jenkins 使用中如何处理 Jenkinsfile 及 Jenkinsfile 的编写方式。

1. 插入字符串

Jenkins 使用与 Groovy 相同的规则进行字符串赋值,可以使用单引号或者双引号进行赋值。例如:

```
def singlyQuoted = 'Hello'
def doublyQuoted = "World"
```

引用变量需要使用双引号:

```
def username = 'Jenkins'
echo 'Hello Mr. ${username}'
echo "I said, Hello Mr. ${username}"
```

运行结果:

```
Hello Mr. ${username}
I said, Hello Mr. Jenkins
```

2. 使用环境变量

Jenkins 有许多内置变量可以直接在 Jenkinsfile 中使用,比如:

- BUILD_ID:当前构建的 ID,与 Jenkins 版本 1.597+中的 BUILD_NUMBER 完全相同。
- JOB_NAME:本次构建的项目名称。
- JENKINS_URL:Jenkins 完整的 URL,需要在 System Configuration 设置。

使用 env.BUILD_ID 和 env.JENKINS_URL 引用内置变量:

```
Jenkinsfile (Declarative Pipeline)
pipeline {
    agent any
    stages {
        stage('Example') {
            steps {
                echo "Running ${env.BUILD_ID} on ${env.JENKINS_URL}"
            }
        }
    }
}
```

对应的脚本式流水线如下:

```
Jenkinsfile (Scripted Pipeline)
node {
    echo "Running ${env.BUILD_ID} on ${env.JENKINS_URL}"
}
```

更多参数请参考以下网址:

https://wiki.jenkins.io/display/JENKINS/Building+a+software+project#Buildingasoftwareproject-JenkinsSetEnvironmentVariables

3. 处理凭证

本节主要介绍在 Jenkins 的使用过程中对一些机密文件的处理方式。

（1）机密文件、用户名密码、私密文件

Jenkins 的声明式流水线语法有一个 credentials() 函数，它支持 secrettext（机密文件）、usernameandpassword（用户名和密码）以及 secretfile（私密文件）。

①机密文件示例

本实例演示将两个 Secret 文本凭证分配给单独的环境变量来访问 Amazon Web 服务，需要提前创建这两个文件的 credentials（下面章节会有演示），Jenkinsfile 文件的内容如下：

```
Jenkinsfile (Declarative Pipeline)
pipeline {
   agent {
      // Define agent details here
   }
   environment {
      AWS_ACCESS_KEY_ID     = credentials('jenkins-aws-secret-key-id')
      AWS_SECRET_ACCESS_KEY = credentials('jenkins-aws-secret-access-key')
   }
   stages {
      stage('Example stage 1') {
         steps {
            //
         }
      }
      stage('Example stage 2') {
         steps {
            //
         }
      }
   }
}
```

说明：

上述示例定义了两个全局变量 AWS_ACCESS_KEY_ID 和 AWS_SECRET_ACCESS_KEY，这两个变量引用的是 credentials 的两个文件，并且这两个变量均可以在 stages 直接引用（$AWS_SECRET_ACCESS_KEY 和$AWS_ACCESS_KEY_ID）。

> **注　意**
>
> 如果在 steps 中使用 echo $AWS_ACCESS_KEY_ID，此时返回的是****，加密内容不会被显示出来。

②用户名密码

本示例用来演示 credentials 账号密码的使用，比如使用一个公用账户访问 Bitbucket、GitLab、Harbor 等，假设已经配置完成了用户名密码的 credentials，凭证 ID 为 jenkins-bitbucket-common-creds。

可以用以下方式设置凭证环境变量：

```
environment {
   BITBUCKET_COMMON_CREDS = credentials('jenkins-bitbucket-common-creds')
}
```

这里实际设置了下面的 3 个环境变量：

- BITBUCKET_COMMON_CREDS，包含一个以冒号分隔的用户名和密码，格式为 username:password。
- BITBUCKET_COMMON_CREDS_USR，仅包含用户名的附加变量。
- BITBUCKET_COMMON_CREDS_PSW，仅包含密码的附加变量。

此时，调用用户名密码的 Jenkinsfile 如下：

```
Jenkinsfile (Declarative Pipeline)
pipeline {
    agent {
        // Define agent details here
    }
    stages {
        stage('Example stage 1') {
            environment {
                BITBUCKET_COMMON_CREDS = credentials('jenkins-bitbucket-common-creds')
            }
            steps {
                //
            }
        }
        stage('Example stage 2') {
            steps {
                //
            }
        }
    }
}
```

> **注　意**
>
> 此时环境变量的凭证仅作用于 stage 1。

4. 处理参数

声明式流水线的参数支持开箱即用，允许流水线在运行时通过 parametersdirective 接受用户指定的参数。

如果将流水线配置为 Build_With_Parameters 选项用来接受参数，那么这些参数将会作为 params 变量被成员访问。假设在 Jenkinsfile 中配置了名为 Greeting 的字符串参数，可以通过 ${params.Greeting} 访问该参数，比如：

```
Jenkinsfile (Declarative Pipeline)
pipeline {
    agent any
    parameters {
        string(name: 'Greeting', defaultValue: 'Hello', description: 'How should I greet the world?')
    }
    stages {
        stage('Example') {
```

```
        steps {
            echo "${params.Greeting} World!"
        }
    }
}
```

对应的脚本式流水线如下：

```
Jenkinsfile (Scripted Pipeline)
properties([parameters([string(defaultValue: 'Hello', description: 'How should I greet the world?', name: 'Greeting')])])

node {
    echo "${params.Greeting} World!"
}
```

5．处理失败

声明式流水线默认通过 postsection 支持健壮的失败处理方式，允许声明许多不同的 postconditions，比如 always、unstable、success、failure 和 changed，具体可参考 Pipeline 的语法。

比如，以下是构建失败发送邮件通知的示例：

```
Jenkinsfile (Declarative Pipeline)
pipeline {
    agent any
    stages {
        stage('Test') {
            steps {
                sh 'make check'
            }
        }
    }
    post {
        always {
            junit '**/target/*.xml'
        }
        failure {
            mail to: team@example.com, subject: 'The Pipeline failed :('
        }
    }
}
```

对应的脚本式流水线如下：

```
Jenkinsfile (Scripted Pipeline)
node {
    /* .. snip .. */
    stage('Test') {
        try {
            sh 'make check'
        }
        finally {
            junit '**/target/*.xml'
        }
    }
}
```

```
        /* .. snip .. */
}
```

6. 使用多个代理

流水线允许在 Jenkins 环境中使用多个代理，这有助于更高级的用例，例如跨多个平台执行构建、测试等。

比如，在 Linux 和 Windows 系统的不同 agent 上进行测试：

```
Jenkinsfile (Declarative Pipeline)
pipeline {
    agent none
    stages {
        stage('Build') {
            agent any
            steps {
                checkout scm
                sh 'make'
                stash includes: '**/target/*.jar', name: 'app'
            }
        }
        stage('Test on Linux') {
            agent {
                label 'linux'
            }
            steps {
                unstash 'app'
                sh 'make check'
            }
            post {
                always {
                    junit '**/target/*.xml'
                }
            }
        }
        stage('Test on Windows') {
            agent {
                label 'windows'
            }
            steps {
                unstash 'app'
                bat 'make check'
            }
            post {
                always {
                    junit '**/target/*.xml'
                }
            }
        }
    }
}
```

4.5 GitLab+ Jenkins +Harbor+ Kubernetes 集成应用

本节介绍持续集成、持续部署的步骤及过程，主要讲解 Jenkins 对应的插件安装、任务（Job）的配置方式和一些基本配置的使用。

4.5.1 基本概念

在 Kubernetes 中使用 CI/CD，一般的步骤为：

（1）在 GitLab 创建对应的项目。
（2）开发者将代码提交到 GitLab。
（3）Jenkins 创建对应的任务（Job），集成该项目的 Git 地址和 Kubernetes 集群。
（4）如有配置钩子，推送（Push）代码会自动触发 Jenkins 构建，如没有配置钩子，需要手动构建。
（5）Jenkins 控制 Kubernetes（使用的是 Kubernetes 插件）创建 Jenkins Slave。
（6）Jenkins Slave 根据流水线（Pipeline）定义的步骤执行构建。
（7）通过 Dockerfile 生成镜像。
（8）将镜像提送（Push）到私有 Harbor。
（9）Jenkins 再次控制 Kubernetes 进行最新的镜像部署。

上面所述为一般步骤，中间还可能会涉及自动化测试等步骤，可自行根据业务场景添加。上面流水线步骤一般写在 Jenkinsfile 中，Jenkins 会自动读取该文件，同时 Jenkinsfile 和 Dockerfile 可一并和代码放置于 GitLab 中，或者单独配置。

4.5.2 基本配置

首先进行持久化安装 Jenkins，请见 3.6 节的内容，然后执行以下步骤。

1. Kubernetes 插件安装

因为使用流水线（Pipeline）进行编译构建，所以只需要安装 Kubernetes Plugin 插件即可。
安装步骤：
首页→系统管理→管理插件→可选插件→Kubernetes plugin 安装。

2. Jenkins Kubernetes 配置

首先需要在 Jenkins 配置 Kubernetes，单击"系统管理"选项，如图 4-1 所示。

图 4-1　配置 Kubernetes

单击页面中的"系统设置"选项，在"系统设置"页面的最下面，用鼠标单击"新增一个云"按钮，然后填写 Kubernetes 的相关信息，如图 4-2 所示。

图 4-2　Kubernetes 基本配置

由于使用的是 Pipeline 进行构建，构建过程中会使用 Jenkinsfile 中指定的容器进行构建，所以上述配置无须填写过多内容，只需要将/root/.kube/config 挂载到 Jenkins Master 的相应目录下即可。可以通过 Secret 挂载，挂载步骤参考 2.2.10 节。

挂载 config 文件后，单击 Test Connection 即可连接，如图 4-3 所示。

第 4 章　持续集成与持续部署　│　235

图 4-3　Kubernetes 链接测试

4.5.3　新建任务（Job）

打开 Jenkins 的首页，单击"新建任务"选项，如图 4-4 所示。

图 4-4　新建任务

填写任务的基本信息，如图 4-5 所示。

图 4-5　任务的基本信息

在任务页面上配置 GitLab 的项目地址，如图 4-6 所示。

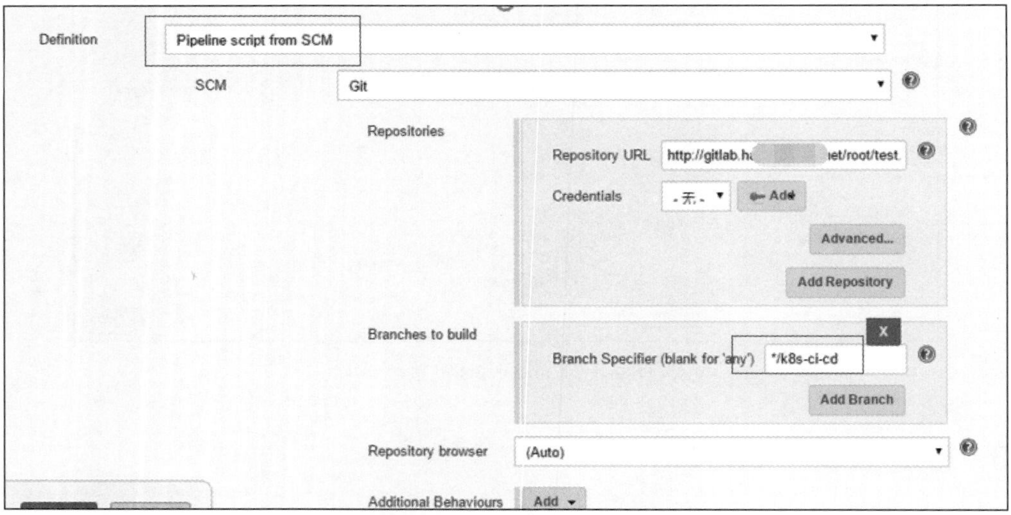

图 4-6　配置 GitLab

选择 Pipeline 脚本，在这里使用的是名称为 Jenkinsfile 的 Jenkinsfile，可以自定义名字和路径，当前为项目的根路径，如图 4-7 所示。

图 4-7　选择 Pipeline 脚本

所有的项目均按照此步骤配置即可，创建完成后单击构建按钮，就会按照 Jenkinsfile 的步骤进行构建。下面一节（4.6 节）将演示具体的构建方法。

4.5.4　Jenkins 凭据的使用

在生产环境中，代码仓库或者镜像仓库都必须设置账户密码或密钥进行认证才能访问，此时可以使用 Jenkins 配置凭据，之后就可以在 Jenkinsfile 中引用镜像仓库的凭据了。同样，在任务（Job）配置中也可以选择代码仓库的 Key。

配置凭据，如图 4-8 所示。

填写凭据变量，如图 4-9 所示。

图 4-8　配置凭据

图 4-9　填写凭据变量

保存后，单击"添加凭据"选项，如图 4-10 所示。

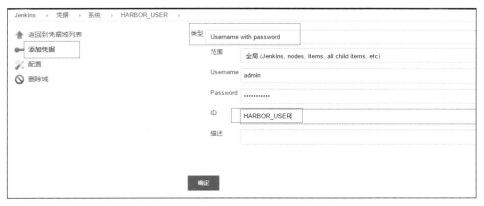

图 4-10　添加凭据

此时，创建了访问 Harbor 的账户密码，GitLab 的密钥配置类似。

然后，可以在 Jenkinsfile 中引用该凭据，比如：

```
stage('Create Docker images') {
    container('docker') {
      withCredentials([[$class: 'UsernamePasswordMultiBinding',
        credentialsId: 'HARBOR_USER',
        usernameVariable: 'HARBOR_USER',
        passwordVariable: 'HARBOR_PASSWORD']]) {
        sh """
          echo '$HARBOR_USER $HARBOR_PASSWORD'
          docker login -u ${HARBOR_USER} -p ${HARBOR_PASSWORD} harbor.xxx.net
          docker build -t harbor.xxx.net/K8S/test:${gitCommit} .
          docker push harbor.xxx.net/K8S/test:${gitCommit}
          """
```

```
        }
      }
    }
```

credentialsId: 'HARBOR_USER'为上述创建的 ID。

4.6 自动化构建 Java 应用

本节演示自动构建 Java 应用，使用的是 Tomcat。由于步骤是按照公司内部代码编写的，因此请将 Dockerfile 和 Jenkinsfile 放置于目标公司代码的根目录下，然后进行测试，或者使用其他项目测试。本示例只是为了演示相关步骤和流水线（Pipeline）的编写。

4.6.1 定义 Dockerfile

Dockerfile 主要写的是如何生成公司业务的镜像：

```
# version 1.0
FROM tomcat:8-jre8-alpine
MAINTAINER xxx "xxx@xxx.net"

RUN rm -rf /usr/local/tomcat/webapps/* && mkdir /usr/local/tomcat/webapps/ROOT -p

COPY ./ROOT /usr/local/tomcat/webapps/ROOT

CMD [ "sh", "-c", "catalina.sh run" ]
```

说明：

首先用 FROM 选择基础镜像，然后通过 RUN 创建项目的目录，再通过 COPY 将编译解压缩后的文件导入到容器中，最后定义 Tomcat 实例在前台启动。

如果公司使用的是微服务，并且通过 java -jar 启动项目，则可以使用 Java 的基础镜像，最后执行命令 java -jar JAR_PACKAGE 即可，当然也可以将启动命令写在部署文件中。

4.6.2 定义 Jenkinsfile

Jenkinsfile 用于定义流水线执行的步骤，此时使用的是 Kubernetes 插件的 podTemplate 编写的 Jenkinsfile，也可以使用上述讲解的声明式流水线进行编写：

```
def label = "worker-${UUID.randomUUID().toString()}"
podTemplate(label: label, containers: [
    containerTemplate(name: 'docker', image: 'docker', command: 'cat', ttyEnabled: true),
    containerTemplate(name: 'maven', image: 'maven:3.5.3-jdk-8-alpine', command: 'cat', ttyEnabled: true),
    containerTemplate(name: 'kubectl', image: 'roffe/kubectl:v1.11.1', command: 'cat', ttyEnabled: true)
  ],
```

```groovy
      volumes: [
        hostPathVolume(mountPath: '/var/run/docker.sock', hostPath:
'/var/run/docker.sock'),
        hostPathVolume(mountPath: '/root/.kube/config', hostPath:
'/root/.kube/config'),
    ]) {
      node(label) {
         def myRepo = checkout scm
         def gitCommit = myRepo.GIT_COMMIT
         def gitBranch = myRepo.GIT_BRANCH
         def shortGitCommit = "${gitCommit[0..10]}"
         def previousGitCommit = sh(script: "git rev-parse ${gitCommit}~",
returnStdout: true)
         def BUILD_NUMBER= "${BUILD_NUMBER}"
         def JOB_NAME = "${JOB_NAME}"
    def NS = "test-java"
    def PROJECT = "trustee-task"
         stage('build') {
           try {
             container('maven') {
               sh """
                 mkdir ROOT -p
                 mvn clean install
                 cd ROOT && jar -xvf ../*/target/*.war
               """
             }
           }
           catch (exc) {
             println "Failed to build - ${currentBuild.fullDisplayName}"
             throw(exc)
           }
         }
         stage('Create Docker images') {
           container('docker') {
             withCredentials([[$class: 'UsernamePasswordMultiBinding',
               credentialsId: 'HARBOR_USER',
               usernameVariable: 'HARBOR_USER',
               passwordVariable: 'HARBOR_PASSWORD']]) {
               sh """
                 docker login -u ${HARBOR_USER} -p ${HARBOR_PASSWORD} harbor.K8S.net
                 docker build -t harbor.K8S.net/${NS}/${JOB_NAME}:${BUILD_NUMBER} .
                 docker tag harbor.K8S.net/${NS}/${JOB_NAME}:${gitCommit}
harbor.K8S.net/${NS}/${JOB_NAME}:latest
                 docker push harbor.K8S.net/${NS}/${JOB_NAME}:${BUILD_NUMBER}
                 docker push harbor.K8S.net/${NS}/${JOB_NAME}:latest
                 rm -rf ROOT
               """
             }
           }
         }
         stage('Run kubectl') {
           container('kubectl') {
             sh """
               # cat /root/.kube/config
               export KUBECONFIG=/root/.kube/config
```

```
                    kubectl set image deployment/${PROJECT}
${PROJECT}=harbor.K8S.net/${NS}/${JOB_NAME}:${BUILD_NUMBER} -n ${NS} --record
                    kubectl get po -n ${NS}
                    kubectl describe deploy ${PROJECT} -n ${NS} | grep Image
                    """
        }
      }
    }
  }
```

说明：

上面定义了 3 个用于构建的 docker，因为该测试项目用 Maven 进行打包，所以使用的是 Maven 的镜像，请按需修改。

本示例镜像仓库按照命名空间进行划分，镜像名称指定的为 Jenkins 的任务（Job）名称，可以直接通过$JOB_NAME 引用，请按需修改。

PROJECT 为 Deployment 的名称，请按需修改。

Kubectl 的版本需要和集群对应，请按需修改。

4.6.3 定义 Deployment

由于使用的是私有仓库，因此需要先配置拉取私有仓库镜像的密钥：

```
[root@K8S-master01 ~]# kubectl create ns test-java
namespace/test-java created
[root@K8S-master01 ~]# kubectl create secret docker-registry harbor
--docker-server=harbor.K8S.net --docker-username=admin
--docker-email=xxx@xxx.com --docker-password=Harbor12345 -n test-java
secret/harbor created
```

说明：test-java 为该示例的命名空间。

预定义 Deployment 之后可以直接使用 Jenkinsfile 中的步骤（Run kubectl）更新镜像：

```
apiVersion: extensions/v1beta1
kind: Deployment
metadata:
  namespace: test-java
  name:trustee-task
  labels:
    app: trustee-task
    env: test
spec:
  replicas: 4
  strategy:
    type: RollingUpdate
    rollingUpdate:
      maxUnavailable: 1
      maxSurge: 2
  minReadySeconds: 5
  template:
    metadata:
      labels:
```

```
      app: trustee-task
spec:
  containers:
    - name: trustee-task
      image: dotbalo/canary:v1
      imagePullPolicy: IfNotPresent
      ports:
        - containerPort: 8080
          name: tomcat
      livenessProbe:
        tcpSocket:
          port: tomcat
        initialDelaySeconds: 20
        periodSeconds: 10
        failureThreshold: 10
      readinessProbe:
        tcpSocket:
          port: tomcat
        initialDelaySeconds: 20
        periodSeconds: 10
        failureThreshold: 10
  imagePullSecrets:
    - name: harbor
```

创建此 Deployment：

```
[root@K8S-master01 4.6.3]# kubectl create -f test-java.yaml
deployment.extensions/trustee-task created
```

4.6.4 Harbor 项目创建

需要在 Harbor 上提前创建该任务（Job）存储镜像的项目（Project），本示例命名为 test-java。

通过浏览器访问 Harbor，并创建 test-java 项目，依次单击"项目"→"新建项目"→"项目名称"，填入项目名称，如图 4-11 所示。

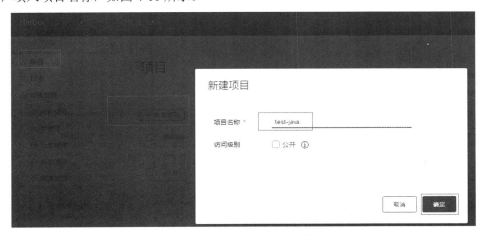

图 4-11　在 Harbor 上创建项目

4.6.5　创建任务（Job）

单击首页的创建任务（Job）选项并配置任务信息，如图 4-12 所示。

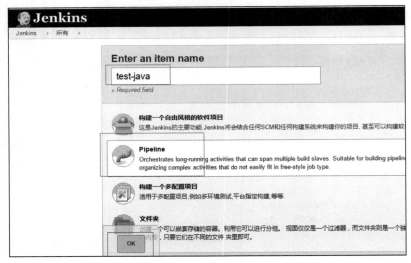

图 4-12　配置任务

配置 Pipeline，如图 4-13 所示。

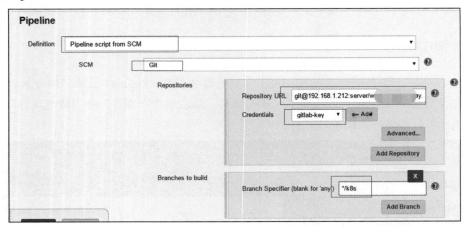

图 4-13　配置 Pipeline

单击图 4-13 中的 Add 按钮，添加 GitLab Key，之后将 sshkey 添加到 key 框中，如图 4-14 所示。

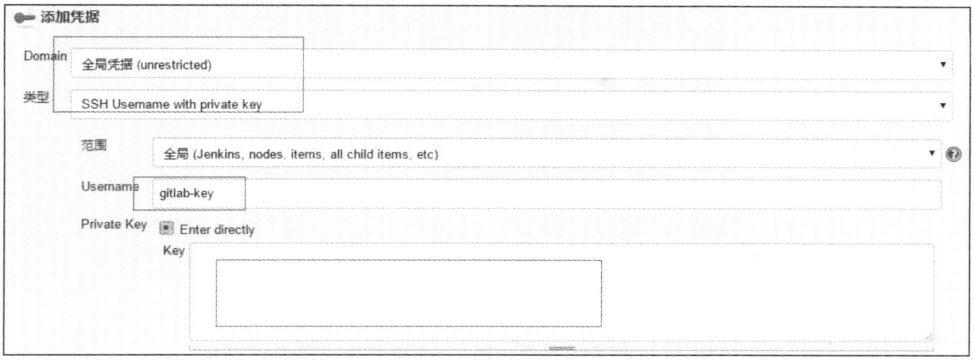

第 4 章 持续集成与持续部署

图 4-14 配置 GitLab Key

最后单击保存任务即可。

4.6.6 执行构建

单击主页选择对应的任务（Job），然后单击"立即构建"选项，如图 4-15 所示。

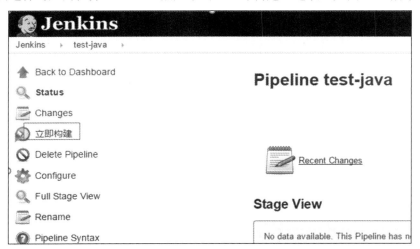

图 4-15 单击"立即构建"

在控制台可以看到相应的输出，如图 4-16 所示。

图 4-16 构建日志

同时，也可以看到流水线（Pipeline）执行的步骤，如图 4-17 所示。

```
Commit message: "test-java"
First time build. Skipping changelog.
[Pipeline] sh
+ git rev-parse '36b871375802fb1d74c0a585b4052e672bcc5766~'
[Pipeline] stage
[Pipeline] { (build)
[Pipeline] container
[Pipeline] {
[Pipeline] sh
+ pwd
/home/jenkins/workspace/test-java
+ ls -a
.
..
.git
.gitignore
Dockerfile
Jenkinsfile
README.md
```

图 4-17　流水线（Pipeline）执行的步骤

构建成功后的输出如图 4-18 所示。

```
[INFO] Reactor Summary:
[INFO]
[INFO] ▓▓▓▓▓▓▓▓ 1.1.6-SNAPSHOT ................. SUCCESS [ 58.236 s]
[INFO] trustee-core ........................... SUCCESS [05:25 min]
[INFO] trustee-domain ......................... SUCCESS [  2.649 s]
[INFO] trustee-task 1.1.6-SNAPSHOT ............ SUCCESS [02:28 min]
[INFO] ------------------------------------------------------------
[INFO] BUILD SUCCESS
[INFO] ------------------------------------------------------------
[INFO] Total time: 08:56 min
[INFO] Finished at: 2019-02-22T15:01:15Z
[INFO] ------------------------------------------------------------
+ cd ROOT
+ jar -xvf ../trustee-task/target/trustee-task.war
  created: META-INF/
 inflated: META-INF/MANIFEST.MF
```

图 4-18　构建成功后的日志输出

镜像构建步骤，如图 4-19 所示。

```
f41fe1b6eee3: Pull complete
65369fd9d03b: Pull complete
7391503a863e: Pull complete
18d5a6a9e618: Pull complete
Digest: sha256:c44fba6b5d74732fe530ac4d64aad8bce4e9cebf4263c542e368d0845385bfdc
Status: Downloaded newer image for tomcat:8-jre8-alpine
 ---> d9257e5204db
Step 2/5 : MAINTAINER Dukuan "dukuan@haixiangjinfu.net"
 ---> Running in 3269eea99cd9
 ---> 6c13697f3afe
Removing intermediate container 3269eea99cd9
Step 3/5 : RUN rm -rf /usr/local/tomcat/webapps/* && mkdir /usr/local/tomcat/webapps/ROOT -p
 ---> Running in fd04d8f2e5de
 ---> 67a4d4b1f437
Removing intermediate container fd04d8f2e5de
Step 4/5 : COPY ./ROOT /usr/local/tomcat/webapps/ROOT
 ---> babab513f9ed
Removing intermediate container 5021a07610ed
Step 5/5 : CMD sh -c catalina.sh run
 ---> Running in a3e07dd290cb
 ---> 31412041254a
Removing intermediate container a3e07dd290cb
Successfully built 31412041254a
+ docker tag harbor.k8s.net/test-java/test-java:9 harbor.k8s.net/test-java/test-java:latest
```

图 4-19　镜像构建步骤

镜像上传步骤，如图 4-20 所示。

```
+ docker push harbor.k8s.net/test-java/test-java:9
The push refers to a repository [harbor.k8s.net/test-java/test-java]
040c8c283e52: Preparing
dc65c2b858d1: Preparing
272341f389fc: Preparing
c4f9fbfc5ef3: Preparing
ca7b7428be5a: Preparing
2acbc8fc5e4d: Preparing
744b4cd8cf79: Preparing
503e53e365f3: Preparing
2acbc8fc5e4d: Waiting
503e53e365f3: Waiting
744b4cd8cf79: Waiting
272341f389fc: Pushed
dc65c2b858d1: Pushed
ca7b7428be5a: Pushed
040c8c283e52: Pushed
744b4cd8cf79: Pushed
503e53e365f3: Pushed
c4f9fbfc5ef3: Pushed
2acbc8fc5e4d: Pushed
9: digest: sha256:b7692db45cbc403304907dfc8b9248bfd36e06221f499bb3ca256943d095925f size: 1993
```

图 4-20　镜像上传步骤

此时，在 Harbor 上可以看到推送（Push）的镜像，如图 4-21 所示。

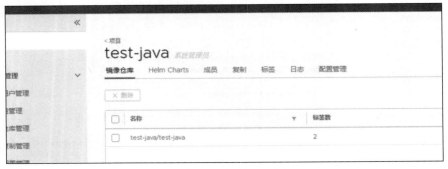

图 4-21　查看 Harbor 镜像

查看镜像详情，如图 4-22 所示。

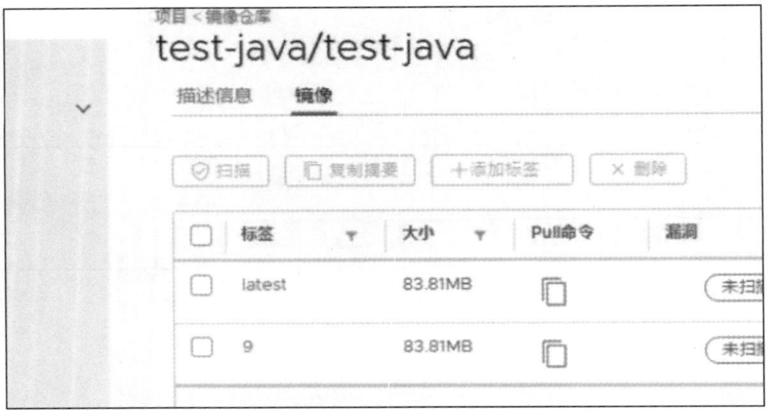

图 4-22　镜像详细信息

此时镜像的 tag 用 BUILD_NUMBER 代替，即任务（Job）的第 N 次构建，可按需修改 tag。最后可以看到 Deployment 的镜像已经被修改，参考图 4-23。

```
+ kubectl set image deployment/trustee-task 'trustee-task=harbor.k8s.net/test-java/test-java:9' -n t
deployment.extensions/trustee-task image updated
+ kubectl get po -n test-java
NAME                            READY   STATUS        RESTARTS   AGE
test-java-9b7468744-6s98k       1/1     Running       0          5h
test-java-9b7468744-9wfwr       1/1     Running       0          5h
test-java-9b7468744-cgf6n       1/1     Running       0          5h
test-java-9b7468744-msnqp       1/1     Running       0          5h
trustee-task-5698db5f8f-8wrt7   0/1     Pending       0          0s
trustee-task-7bdc59f49d-7gxlv   1/1     Running       0          35m
trustee-task-7bdc59f49d-b4vb4   1/1     Running       0          35m
trustee-task-7bdc59f49d-s74mh   1/1     Terminating   0          35m
trustee-task-7bdc59f49d-t79w2   1/1     Running       0          35m
+ kubectl describe deploy trustee-task -n test-java
+ grep Image
    Image:      harbor.k8s.net/test-java/test-java:9
[Pipeline] }
[Pipeline] // container
[Pipeline] }
[Pipeline] // stage
[Pipeline] }
[Pipeline] // node
[Pipeline] }
[Pipeline] // podTemplate
[Pipeline] End of Pipeline
Finished: SUCCESS
```

图 4-23　Deployment 更新

本实例演示的代码为笔者公司内部代码，不方便公开，读者可以使用自己公司的代码进行测试。本节主要讲解的是持续部署的一般步骤，可按照上面的演示方式更改为适合自己的流水线。

4.7 自动化构建 NodeJS 应用

本节介绍自动化构建 NodeJS 应用，其构建方式和自动化构建 Java 基本相同，重点是更改 Deployment、Jenkinsfile 和 Dockerfile。

4.7.1 定义 Dockerfile

如果 NodeJS 仅仅作为前端，在使用 NPM 进行编译后，一般在 dist 目录会生成相应的 html 文件，可以直接使用 Nginx 进行部署：

```
# version 1.0
FROM nginx:stable-alpine
MAINTAINER xxx"xxx@xxx.net"

COPY ./dist /usr/share/nginx/html

# CMD [ "sh", "-c", "catalina.sh run" ]
ENTRYPOINT [ "/usr/sbin/nginx", "-g", "daemon off;" ]
```

此文件使用的是 nginx:stable-alpine 作为基础镜像，之后将 dist 的 html 文件拷贝到 Nginx 的 /usr/share/nginx/html 目录中，再启动 Nginx 即可访问。

4.7.2 定义 Deployment

```
apiVersion: extensions/v1beta1
kind: Deployment
metadata:
  namespace: test-nodejs
  name: test-nodejs
  labels:
    app: test-nodejs
    env: test
spec:
  replicas: 4
  strategy:
    type: RollingUpdate
    rollingUpdate:
      maxUnavailable: 1
      maxSurge: 2
  minReadySeconds: 5
  template:
    metadata:
      labels:
        app: test-nodejs
```

```yaml
    spec:
      containers:
        - name: test-nodejs
          image: harbor.K8S.net/test-nodejs/test-nodejs:f17023f14ebbe33d8c7394f76bc8031600e04799
          imagePullPolicy: IfNotPresent
          ports:
            - containerPort: 80
              name: wap
          livenessProbe:
            tcpSocket:
              port: wap
            initialDelaySeconds: 20
            periodSeconds: 10
            failureThreshold: 10
          readinessProbe:
            tcpSocket:
              port: wap
            initialDelaySeconds: 20
            periodSeconds: 10
            failureThreshold: 10
      imagePullSecrets:
        - name: harbor
```

4.7.3 定义 Jenkinsfile

```groovy
def label = "worker-${UUID.randomUUID().toString()}"

podTemplate(label: label, containers: [
    containerTemplate(name: 'docker', image: 'docker', command: 'cat', ttyEnabled: true),
    containerTemplate(name: 'node6', image: 'node:6.15.1-alpine', command: 'cat', ttyEnabled: true),
    containerTemplate(name: 'kubectl', image: 'roffe/kubectl:v1.11.1', command: 'cat', ttyEnabled: true)
],
volumes: [
    hostPathVolume(mountPath: '/var/run/docker.sock', hostPath: '/var/run/docker.sock'),
    hostPathVolume(mountPath: '/root/.kube/config', hostPath: '/root/.kube/config')
]) {
    node(label) {
        def myRepo = checkout scm
        def gitCommit = myRepo.GIT_COMMIT
        def gitBranch = myRepo.GIT_BRANCH
        def shortGitCommit = "${gitCommit[0..10]}"
        def previousGitCommit = sh(script: "git rev-parse ${gitCommit}~", returnStdout: true)
        def JOB_NAME = "test-nodejs"
    def PROJECT = "test-nodejs"
    def NS = "test-nodejs"
        stage('Exec NPM INSTALL or no...') {
            try {
```

```
          container('node6') {
              sh """
                npm install
              """
          }
      }
      catch (exc) {
        println "Failed to INSTALL - ${currentBuild.fullDisplayName}"
        throw(exc)
      }
    }

    stage('Run build') {
      container('node6') {
        sh """
          npm run build
        """
      }
    }

    stage('Create Docker images') {
      container('docker') {
        withCredentials([[$class: 'UsernamePasswordMultiBinding',
            credentialsId: 'HARBOR_USER',
            usernameVariable: 'HARBOR_USER',
            passwordVariable: 'HARBOR_PASSWORD']]) {
          sh """
            docker login -u ${HARBOR_USER} -p ${HARBOR_PASSWORD} harbor.K8S.net
            docker build -t harbor.K8S.net/${NS}/${PROJECT}:${gitCommit} .
            docker tag harbor.K8S.net/${NS}/${PROJECT}:${gitCommit} harbor.K8S.net/${PROJECT}/${PROJECT}:latest
            docker push harbor.K8S.net/${NS}/${PROJECT}:${gitCommit}
            docker push harbor.K8S.net/${NS}/${PROJECT}:latest
          """
        }
      }
    }
    stage('Run kubectl') {
      container('kubectl') {
        sh """
          # cat /root/.kube/config
          export KUBECONFIG=/root/.kube/config
          kubectl set image deployment/${PROJECT} ${PROJECT}=harbor.K8S.net/${NS}/${JOB_NAME}:${gitCommit} -n ${NS} --record
        """
      }
    }
  }
}
```

这里 Jenkinsfile 中的镜像（image）的 tag 是按照 Git Commit 的值进行定义的，tag 可按照自己的方式定义，没有必要全部只按照一种方式定义，请按需定义。

其他步骤和自动化构建 Java 类似，可按照上一节的步骤配置。

4.8 自动化构建 Spring Cloud 应用

Spring Cloud 是基于 Spring Boot 的一整套实现微服务的框架，目前已经被很多公司用于快速开发及迭代业务应用。Spring Cloud 可将公司大项目拆分成多个独立功能的微服务，减少了升级某个业务功能对整个业务的影响。Spring Cloud 提供了开发所需的分布式会话、配置管理、服务发现、断路器、智能路由、控制总线和全局锁等组件。

Spring Cloud 常用的子项目如下。

- Spring Cloud Config：配置管理工具包，可以将配置统一放置到远程服务器上进行集中化管理集群及配置，目前支持本地存储、Git 和 Subversion。
- Spring Cloud Bus：事件、消息总线，用于集群中传播状态变化，可与 Spring Cloud Config 联合实现热部署。
- Eureka：云端服务发现、注册中心，一个基于 REST 的服务，其他应用注册到 Eureka，以实现中间层服务发现和故障转移。
- Hystrix：熔断器，容错管理工具，旨在通过熔断机制控制服务和第三方库的节点，从而对延迟和故障提供更强大的容错能力。
- Zuul：Zuul 是在云平台上提供动态路由、监控、弹性、安全等边缘服务的框架。Zuul 相当于是设备和 Netflix 流应用的 Web 网站后端所有请求的前门。
- Consul：封装了 Consul 操作，是一个服务发现与配置工具，可与 Docker 容器无缝集成。

更多详情可以查看：

Spring Cloud 官网文档：https://spring.io/
Spring Cloud 中文文档：https://springcloud.cc/
Spring Cloud 中文社区：http://springcloud.cn/

4.8.1 自动化构建 Eureka

Eureka 作为 Spring Cloud 的服务发现注册中心，其他应用都会注册到 Eureka，在 Eureka 的界面上可以看到每个应用的信息。一般情况下，Eureka 部署三个节点组成集群，本节将使用 StatefulSet 部署 Eureka，每个 Eureka 通过 Headless Service 进行通信。

1. 定义 Eureka 配置文件

在容器化部署 Eureka 时，各节点通信采用的是 Headless Service 的形式。使用 StatefulSet 部署时，StatefulSet 会给每个 Eureka 的 Pod 配置一个 Service，Service 的 FQDN 如下：eureka-x.eureka.NAMESPACE，具体规则可查看 2.2.7 节。我们需要提前将 Eureka 的 defaultZone 改为如下配置，其中 sc 为 eureka 部署的命名空间：

```
defaultZone:
http://eureka-0.eureka.sc:8761/eureka/,http://eureka-1.eureka.sc:8761/eureka/,
```

```
http://eureka-2.eureka.sc:8761/eureka/
```

如果采用的是 Config 配置中心，只需要修改 Config 的统一配置即可，一般此配置由开发人员更改，如果读者对配置中心有所了解也可自行进行修改。

2. 定义 Dockerfile

定义 Dockerfile 的形式和之前并无差别，之前定义的 Dockerfile 应用程序启动被定义到了 Dockerfile 中，这种方式并不是最佳实践，因为这种定义方式在更改启动命令时需要编译镜像，所以本书的自动化部署会将启动命令写于部署文件中。

定义 Dockerfile 如下：

```
# version 1.0
# base-image: java:8u111-jre8-alpine, ap: alpine, sc: Spring Cloud

FROM java:8u111-jre8-alpine
MAINTAINER xxx

# add jar to workdir
COPY target/*.jar /home/tomcat
EXPOSE 8761
```

这里的基础镜像采用官方的 Java 镜像，也可以指定为自己构建的镜像。

> **注 意**
>
> 由 alpine 作为基础镜像做成的 jre 镜像，Java 一般为 openJDK，可能无法直接在业务应用上使用，所以在实际使用时，需要制作适合于公司业务的基础镜像。

3. 定义 StatefulSet

本节部署采用 StatefulSet 的形式，并且将 Eureka 部署在不同的节点上（podAntiAffinity 定义）。注意 Deployment 文件的命名空间为 sc，配置文件的名称为 profilename，配置文件的名称需要与开发人员进行确认，或者为自行修改的配置文件名称。配置中心 Config Server 的地址通过 -Dspring.cloud.config.uri 参数提供，本节演示的 Config Server 也是采用容器化的部署方式，可以参考 4.8.2 节，定义 StatefulSet 如下：

```
apiVersion: apps/v1beta1
kind: StatefulSet
metadata:
  name: eureka
  namespace: sc
spec:
  replicas: 3
  updateStrategy:
    type: RollingUpdate
  serviceName: eureka
  template:
    metadata:
      labels:
        app: eureka
        env: release
```

```yaml
      spec:
        affinity:
          podAntiAffinity:
            requiredDuringSchedulingIgnoredDuringExecution:
              - labelSelector:
                  matchExpressions:
                    - key: "app"
                      operator: In
                      values:
                        - eureka
                topologyKey: "kubernetes.io/hostname"
        containers:
        - name: eureka
          imagePullPolicy: IfNotPresent
          volumeMounts:
            - name: tz-config
              mountPath: /etc/localtime
          image: harbor.K8S.net/test-sc/eureka:latest
          resources:
            requests:
              memory: "2Gi"
              cpu: "1"
            limits:
              cpu: "1"
              memory: "4Gi"
          env:
            - name: TZ
              value: "Asia/Shanghai"
            - name: LANG
              value: "en_US.utf8"
          ports:
          - containerPort: 8761
            name: web-console
          command:
          - sh
          - -c
          - "java -jar -Dspring.cloud.config.uri=http://config:8888 \
    -Dspring.profiles.active=profilename ./*.jar"
          readinessProbe:        # 可选,容器状态检查
            tcpSocket:     # 检测方式
              port: 8761    # 监控端口
            timeoutSeconds: 2     # 超时时间
            initialDelaySeconds: 30    # 初始化时间
          livenessProbe:    # 可选,监控状态检查
            tcpSocket:     # 检测方式
              port: 8761
            initialDelaySeconds: 30    # 初始化时间
            timeoutSeconds: 2     # 超时时间
        volumes:
          - name: tz-config
            hostPath:
              path: /usr/share/zoneinfo/Asia/Shanghai
```

定义 Service 如下:

```yaml
apiVersion: v1
kind: Service
metadata:
  name: eureka
  namespace: sc
  labels:
    app: eureka
spec:
  ports:
  - port: 8761
    name: web-console
  clusterIP: None
  selector:
    app: eureka

---
kind: Service
apiVersion: v1
metadata:
  labels:
    app: eureka
    type: LoadBalancer
  name: eureka-balancer
  namespace: sc
spec:
  ports:
  - name: http
    port: 8761
    targetPort: 8761
    nodePort: 32111
  selector:
    app: eureka
  type: NodePort
```

其中 clusterIP 为 None 的 Service 是 StatefulSet 通信用的 Service，type 为 NodePort 的 Service 是访问 Eureka 界面的 Service。

4. 定义 Jenkinsfile

定义 Jenkinsfile 如下：

```
def label = "worker-${UUID.randomUUID().toString()}"

podTemplate(label: label, containers: [
    containerTemplate(name: 'jnlp', image: 'jnlp-slave:alpine', args: '${computer.jnlpmac} ${computer.name}'),
    containerTemplate(name: 'docker', image: 'docker:18.06', command: 'cat', ttyEnabled: true),
    containerTemplate(name: 'maven', image: 'maven:3.3.9-8u144', command: 'cat', ttyEnabled: true),
    containerTemplate(name: 'kubectl', image: 'kubectl:v1.13.4', command: 'cat', ttyEnabled: true)
  ],
  volumes: [
    hostPathVolume(mountPath: '/var/run/docker.sock', hostPath:
```

```
'/var/run/docker.sock'),
    hostPathVolume(mountPath: '/etc/hosts', hostPath: '/etc/hosts'),
  ]) {
    node(label) {
      def myRepo = checkout scm
      def gitCommit = myRepo.GIT_COMMIT
      def gitBranch = myRepo.GIT_BRANCH
      def shortGitCommit = "${gitCommit[0..10]}"
      def previousGitCommit = sh(script: "git rev-parse ${gitCommit}~",
returnStdout: true)
      def JOB_NAME = "${JOB_NAME}"
      def JOB_NUMBER = "${BUILD_NUMBER}"
      def HARBOR_ADDRESS = "${HARBOR_ADDRESS}"
      def NS = "${NAMESPACE}"
      def APP_IMAGE_URL = "${HARBOR_ADDRESS}/${NS}/${JOB_NAME}"
      stage('build') {
        try {
          container('maven') {
            sh """
              echo '***start to build***'
   mvn clean install
            """
          }
        }
        catch (exc) {
          println "Failed to build - ${currentBuild.fullDisplayName}"
          throw(exc)
        }
      }
      stage('Create Docker images') {
        container('docker') {
          withCredentials([[$class: 'UsernamePasswordMultiBinding',
            credentialsId: 'HARBOR_USER',
            usernameVariable: 'HARBOR_USER',
            passwordVariable: 'HARBOR_PASSWORD']]) {
            sh """
              echo '*****make image*****'
              docker login -u ${HARBOR_USER} -p ${HARBOR_PASSWORD}
${HARBOR_ADDRESS}
              docker build -t ${APP_IMAGE_URL}:${JOB_NUMBER} .
              docker tag ${APP_IMAGE_URL}:${JOB_NUMBER} ${APP_IMAGE_URL}:latest
              docker push ${APP_IMAGE_URL}:${JOB_NUMBER}
              docker push ${APP_IMAGE_URL}:latest
            """
          }
        }
      }
      stage('Run kubectl') {
        container('kubectl') {
          sh """
            echo '***update  deploy***'
            kubectl set image sts/${JOB_NAME}
${JOB_NAME}=${APP_IMAGE_URL}:${JOB_NUMBER} -n ${NS} --record
            kubectl get po -n ${NS} -l app=${JOB_NAME} -w
            echo '***Deploy Finished***'
```

```
                """
            }
        }
    }
}
```

配置几乎和之前的相同，大部分情况下，一个 Jenkinsfile 可以用于很多同类型的项目，这里 Jenkinsfile 将 Harbor 的地址参数化，需要在任务（Job）上添加参数名为 HARBOR_ADDRESS 的变量，并且镜像的 tag 采用 Job Number 进行区分，在实际使用时，可以选择使用 gitcommit 或者 Job Number 作为镜像的 tag。

通过容器化部署 Eureka 至 Kubernetes 集群后，其余 SpringBoot 可以通过以下配置注册至该 Eureka（同一个 NameSpace 只需要写 eureka-X.eureka 即可，不通 NameSpace 需要写成 eureka-X.eureka.NAMESPACE_NAME）：

```yaml
spring:
    profiles: profilename
eureka:
  instance:
    preferIpAddress: true
  client:
    serviceUrl:
      defaultZone: http://eureka-0.eureka:8761/eureka/,http://eureka-1.eureka:8761/eureka/,http://eureka-2.eureka:8761/eureka/
    healthcheck:
      enabled: true
```

4.8.2 自动化构建 Config

Config 作为 Spring Cloud 的统一配置中心，可以对各个应用的配置进行统一管理，在应用启动时通过-Dspring.profiles.active=profilename 参数指定对应的配置文件，即可读取到相应的配置。在实际使用时，Config Server 应该第一个被部署，以便其他应用能成功读取到相对应的配置。

1. 定义 Dockerfile

定义 Config Server 的 Dockerfile 和之前也并无多大区别，Dockerfile 内容如下：

```dockerfile
# version 1.0
# base-image: java:8u111-jre8-alpine, ap: alpine, sc: Spring Cloud
# config-server Dockerfile
FROM java:8u111-jre8-alpine
MAINTAINER xxx

# add jar to workdir
COPY target/*.jar /home/tomcat
EXPOSE 8888
```

这里同样将启动命令放置于部署文件中，便于后期更改。

2. 定义 Deployment

部署 Config Server 和其他业务应用类似，采用 Deployment 即可，定义 Deployment 如下：

```yaml
apiVersion: extensions/v1beta1
kind: Deployment
metadata:
  namespace: sc
  name: config
  labels:
    app: config
    env: release
spec:
  replicas: 1
  strategy:
    type: RollingUpdate
    rollingUpdate:
      maxUnavailable: 0
      maxSurge: 1
  minReadySeconds: 30
  template:
    metadata:
      labels:
        app: config
    spec:
      containers:
        - name: config
          image: harbor.K8S.net/test-sc/config:latest
          resources:
            requests:
              memory: "1Gi"
            limits:
              memory: "2Gi"
          imagePullPolicy: IfNotPresent
          volumeMounts:
            - name: tz-config
              mountPath: /etc/localtime
          env:
            - name: TZ
              value: "Asia/Shanghai"
            - name: LANG
              value: "en_US.utf8"
            - name: MEMORY_LIMIT
              valueFrom:
                resourceFieldRef:
                  resource: requests.memory
                  divisor: 1Mi
          command:
            - sh
            - -c
            - "java -jar -Dspring.profiles.active=prifilename  ./*.jar "
          ports:
            - containerPort: 8888
              name: tomcat
          livenessProbe:
            tcpSocket:
              port: tomcat
            initialDelaySeconds: 20
            timeoutSeconds: 2
```

```yaml
          failureThreshold: 2
        readinessProbe:
          tcpSocket:
            port: tomcat
          initialDelaySeconds: 20
          timeoutSeconds: 2
          failureThreshold: 2
      imagePullSecrets:
        - name: myregistrykey
      volumes:
        - name: tz-config
          hostPath:
            path: /usr/share/zoneinfo/Asia/Shanghai
```

因为 Config Server 需要被其他应用访问，所以需要建一个对应的 Service，之后集群内的应用可以通过 Config 对其访问：

```yaml
apiVersion: v1
kind: Service
metadata:
  name: config
  namespace: pscm
  labels:
    app: config
spec:
  ports:
  - port: 8888
    name: tomcat
  type: ClusterIP
  selector:
    app: config
```

3. 定义 Jenkinsfile

定义 Jenkinsfile 的内容和 Eureka 的内容并无太大区别，只需要将'kubectl'步骤滚动更新的 sts 改成 Deploy 即可。在实际使用中，也可以将此处设置为变量，通过参数化构建来修改对应参数即可：

```groovy
def label = "worker-${UUID.randomUUID().toString()}"

podTemplate(label: label, containers: [
    containerTemplate(name: 'jnlp', image: 'jnlp-slave:alpine', args: '${computer.jnlpmac} ${computer.name}'),
    containerTemplate(name: 'docker', image: 'docker:18.06', command: 'cat', ttyEnabled: true),
    containerTemplate(name: 'maven', image: 'maven:3.3.9-8u144', command: 'cat', ttyEnabled: true),
    containerTemplate(name: 'kubectl', image: 'kubectl:v1.13.4', command: 'cat', ttyEnabled: true)
  ],
  volumes: [
    hostPathVolume(mountPath: '/var/run/docker.sock', hostPath: '/var/run/docker.sock'),
    hostPathVolume(mountPath: '/etc/hosts', hostPath: '/etc/hosts'),
  ]) {
```

```groovy
    node(label) {
      def myRepo = checkout scm
      def gitCommit = myRepo.GIT_COMMIT
      def gitBranch = myRepo.GIT_BRANCH
      def shortGitCommit = "${gitCommit[0..10]}"
      def previousGitCommit = sh(script: "git rev-parse ${gitCommit}~", returnStdout: true)
      def JOB_NAME = "${JOB_NAME}"
      def JOB_NUMBER = "${BUILD_NUMBER}"
      def HARBOR_ADDRESS = "${HARBOR_ADDRESS}"
      def NS = "${NAMESPACE}"
      def APP_IMAGE_URL = "${HARBOR_ADDRESS}/${NS}/${JOB_NAME}"
      stage('build') {
        try {
          container('maven') {
            sh """
              echo '***start to build***'
              mvn clean install
            """
          }
        }
        catch (exc) {
          println "Failed to build - ${currentBuild.fullDisplayName}"
          throw(exc)
        }
      }
      stage('Create Docker images') {
        container('docker') {
          withCredentials([[$class: 'UsernamePasswordMultiBinding',
            credentialsId: 'HARBOR_USER',
            usernameVariable: 'HARBOR_USER',
            passwordVariable: 'HARBOR_PASSWORD']]) {
            sh """
              echo '*****make image*****'
              docker login -u ${HARBOR_USER} -p ${HARBOR_PASSWORD} ${HARBOR_ADDRESS}
              docker build -t ${APP_IMAGE_URL}:${JOB_NUMBER} .
              docker tag ${APP_IMAGE_URL}:${JOB_NUMBER} ${APP_IMAGE_URL}:latest
              docker push ${APP_IMAGE_URL}:${JOB_NUMBER}
              docker push ${APP_IMAGE_URL}:latest
            """
          }
        }
      }
      stage('Run kubectl') {
        container('kubectl') {
          sh """
            echo '***update deploy***'
            kubectl set image deploy/${JOB_NAME} ${JOB_NAME}=${APP_IMAGE_URL}:${JOB_NUMBER} -n ${NS} --record
            kubectl get po -n ${NS} -l app=${JOB_NAME} -w
            echo '***Deploy Finished***'
          """
        }
      }
    }
```

```
    }
  }
```

4.8.3 自动化构建 Zuul

在 Spring Cloud 框架中，Zuul 相当于集群组件的流量入口，流量到达 Zuul 后，Zuul 再将流量分发至对应的应用中。此时，可以使用 DaemonSet 的方式部署 Zuul 至 Spring Cloud 的节点上，当然这不是必须的，可以根据实际场景选择部署方式。

1. 定义 Dockerfile

定义 Dockerfile 和之前的方式相同，只是端口不同，之前构建的镜像采用的端口都是默认的应用端口，在实际使用中可以统一端口，这样就可以使用同一个 Dockerfile 文件进行所有的应用构建，定义 Dockerfile 文件如下：

```
# version 1.0
# base-image: java:8u111-jre8-alpine, ap: alpine, sc: Spring Cloud
# config-server Dockerfile
FROM java:8u111-jre8-alpine
MAINTAINER xxx

# add jar to workdir
COPY target/*.jar /home/tomcat
EXPOSE 19001
```

2. 定义 DaemonSet

这里将 Zuul 通过 DaemonSet 的方式部署于集群中的每个节点，之后将 Spring Cloud 流量指向 Zuul 的 Service 即可访问 Spring Cloud 应用，定义 DaemonSet 如下：

```
kind: DaemonSet
apiVersion: extensions/v1beta1
metadata:
  name: zuul
  namespace: sc
  labels:
    app: zuul
spec:
  updateStrategy:
    rollingUpdate:
      maxUnavailable: 1
    type: RollingUpdate
  template:
    metadata:
      labels:
        app: zuul
        name: zuul
    spec:
      imagePullSecrets:
        - name: myregistrykey
      volumes:
        - name: tz-config
          hostPath:
```

```yaml
        path: /usr/share/zoneinfo/Asia/Shanghai
    containers:
    - image: harbor.K8S.net/test-sc/zuul:latest
      imagePullPolicy: IfNotPresent
      env:
        - name: TZ
          value: "Asia/Shanghai"
        - name: LANG
          value: "en_US.utf8"
        - name: ENV
          value: "k8prelive"
      command:
        - sh
        - -c
        - "java -jar \
              -Dspring.cloud.config.uri=http://config:8888 \
              -Dspring.profiles.active=profilename ./*.jar"
      livenessProbe:
        failureThreshold: 3
        tcpSocket:
          port: 19001
        initialDelaySeconds: 10
        periodSeconds: 10
        successThreshold: 1
        timeoutSeconds: 1
      readinessProbe:
        failureThreshold: 3
        tcpSocket:
          port: 19001
        periodSeconds: 10
        successThreshold: 1
        timeoutSeconds: 1
      volumeMounts:
        - name: tz-config
          mountPath: /etc/localtime
      name: zuul
      ports:
      - name: zuul
        containerPort: 19001
```

要求 Zuul 是 Spring Cloud 的流量入口，所以需要定义 Service 供其他业务访问，Service 的类型请按需进行更改：

```yaml
apiVersion: v1
kind: Service
metadata:
  name: zuul-svc
  namespace: sc
  labels:
    app: zuul
spec:
  type: NodePort
  ports:
  - port: 19001
    name: zuul
```

```
    targetPort: 19001
    nodePort: 31101
  selector:
    app: zuul
```

3. 定义 Jenkinsfile

这里将应用的部署类型也参数化，在 Jenkins Job 下添加对应的参数即可，这样的话，就可以采用统一的 Jenkinsfile 构建所有的 Spring Cloud 应用，定义 Jenkinsfile 如下：

```
    def label = "worker-${UUID.randomUUID().toString()}"

    podTemplate(label: label, containers: [
      containerTemplate(name: 'jnlp', image: 'jnlp-slave:alpine', args: '${computer.jnlpmac} ${computer.name}'),
      containerTemplate(name: 'docker', image: 'docker:18.06', command: 'cat', ttyEnabled: true),
      containerTemplate(name: 'maven', image: 'maven:3.3.9-8u144', command: 'cat', ttyEnabled: true),
      containerTemplate(name: 'kubectl', image: 'kubectl:v1.13.4', command: 'cat', ttyEnabled: true)
    ],
    volumes: [
      hostPathVolume(mountPath: '/var/run/docker.sock', hostPath: '/var/run/docker.sock'),
      hostPathVolume(mountPath: '/etc/hosts', hostPath: '/etc/hosts'),
    ]) {
      node(label) {
        def myRepo = checkout scm
        def gitCommit = myRepo.GIT_COMMIT
        def gitBranch = myRepo.GIT_BRANCH
        def shortGitCommit = "${gitCommit[0..10]}"
        def previousGitCommit = sh(script: "git rev-parse ${gitCommit}~", returnStdout: true)
        def JOB_NAME = "${JOB_NAME}"
        def JOB_NUMBER = "${BUILD_NUMBER}"
        def HARBOR_ADDRESS = "${HARBOR_ADDRESS}"
  def NS = "${NAMESPACE}"
  def DEPLOY_TYPE = "${DEPLOY_TYPE}"
        def APP_IMAGE_URL = "${HARBOR_ADDRESS}/${NS}/${JOB_NAME}"
        stage('build') {
          try {
            container('maven') {
              sh """
                echo '***start to build***'
                mvn clean install
              """
            }
          }
          catch (exc) {
            println "Failed to build - ${currentBuild.fullDisplayName}"
            throw(exc)
          }
        }
        stage('Create Docker images') {
```

```
            container('docker') {
              withCredentials([[$class: 'UsernamePasswordMultiBinding',
                credentialsId: 'HARBOR_USER',
                usernameVariable: 'HARBOR_USER',
                passwordVariable: 'HARBOR_PASSWORD']]) {
                sh """
                  echo '*****make image*****'
                  docker login -u ${HARBOR_USER} -p ${HARBOR_PASSWORD} ${HARBOR_ADDRESS}
                  docker build -t ${APP_IMAGE_URL}:${JOB_NUMBER} .
                  docker tag ${APP_IMAGE_URL}:${JOB_NUMBER} ${APP_IMAGE_URL}:latest
                  docker push ${APP_IMAGE_URL}:${JOB_NUMBER}
                  docker push ${APP_IMAGE_URL}:latest
                """
              }
            }
          }
        stage('Run kubectl') {
          container('kubectl') {
            sh """
              echo '***update deploy***'
              kubectl set image ${DEPLOY_TYPE}/${JOB_NAME} ${JOB_NAME}=${APP_IMAGE_URL}:${JOB_NUMBER} -n ${NS} --record
              kubectl get po -n ${NS} -l app=${JOB_NAME} -w
              echo '***Deploy Finished***'
            """
          }
        }
      }
```

4.9 Webhook 介绍

Webhook 是用于自动化触发 Jenkins 进行构建的工具，一般开发提交代码后，会自动触发 Jenkins 进行构建，不用进入到 Jenkins 中执行构建。

4.9.1 安装 Webhook 插件

使用 Webhook 需要安装的插件如下，如图 4-24 所示。

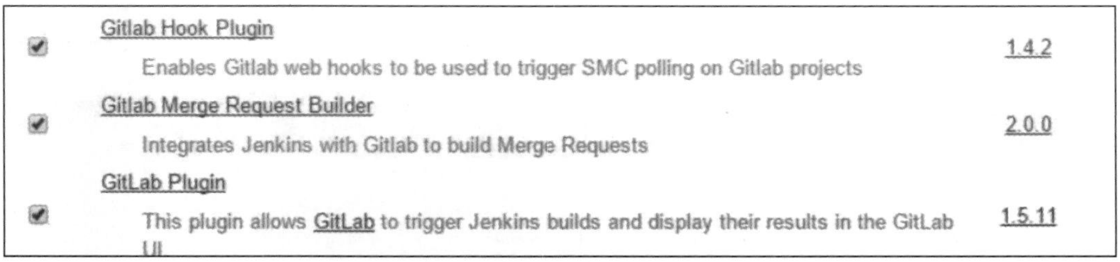

图 4-24　Webhook 插件

4.9.2 配置 Jenkins

安装完 Webhook 以后，需要配置 Jenkins 的任务（Job），选择对应的任务后，单击"配置"按钮，然后勾选 Build when a change is pushed to…，如图 4-25 所示。

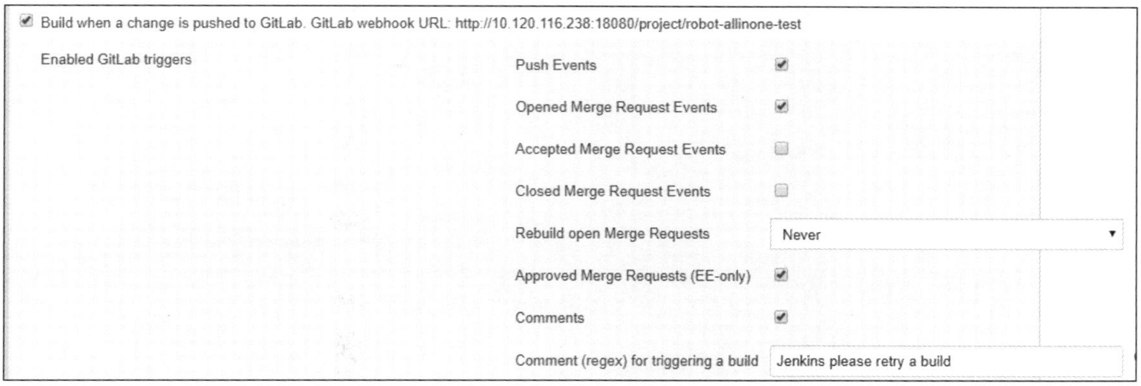

图 4-25　配置 Jenkins

配置自动触发构建的分支，如图 4-26 所示。

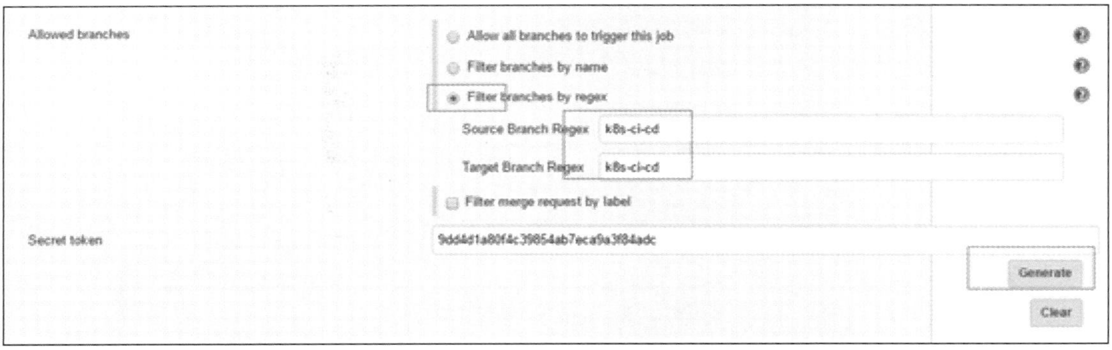

图 4-26　配置分支（此图模糊）

上述配置是在 K8S-ci-cd 分支推送（Push）和合并（Merge）代码后，该配置会触发 Jenkins 进行自动构建，可按需配置。

4.9.3 配置 GitLab

在 GitLab 上启用该 Webhook，如图 4-27 所示。

图 4-27 配置 GitLab（此图模糊）

URL 为 Jenkins Job 中 Build when a change 后面的 URL，如图 4-28 所示。

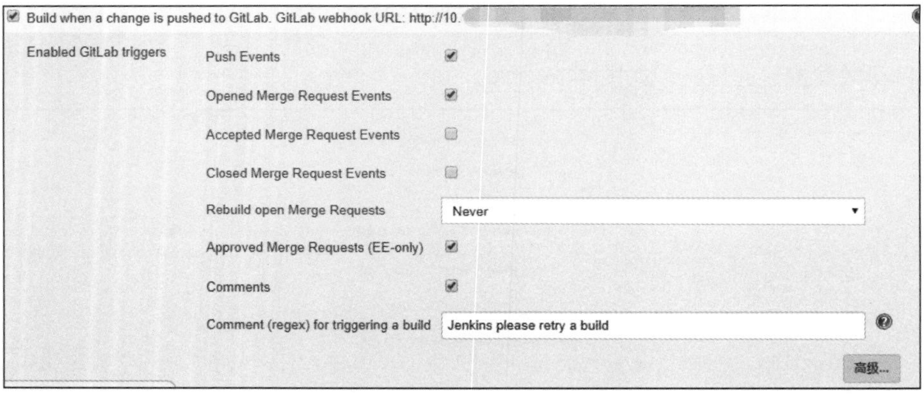

图 4-28 Jenkins 触发 URL

Secret Token 为 Jenkins Job 配置页面的 Secret token（单击 Generate 自动生成），参考图 4-29。

图 4-29 Secret Token

此时推送（Push）代码到 K8S-ci-cd 分支即可触发 Jenkins 进行构建，送其他分支不会触发。

4.10 自动化构建常见问题的解决

虽然上面一节实现了自动化构建应用并发布到 Kubernetes 集群中，但在实际使用过程中，还有许多细节问题需要注意，比如自动化构建过程中代码拉取的速度过慢、Maven 编译的过程过慢、NodeJS 和 PHP 依赖组件安装慢以及 Jenkinsfile 不够灵活等，这些问题在实际使用中都是难以避免的，本小节将对之前的 Jenkins 构建过程进行优化，以提高自动化构建的效率。

4.10.1 解决代码拉取速度慢的问题

在使用 Jenkins 或者其他工具构建的过程中，拉取代码是必不可少的步骤，之前演示的 Jenkins 流水线构建模式是在 Kubernetes 集群中创建 Jenkins Slave Pod 进行构建的，构建完成后 Jenkins Slave Pod 将会被销毁，在构建过程中下载的代码及依赖文件都将会被删除，当下次再次构建的时候会再次下载一次。当项目过于庞大、源代码又比较多时，构建过程中拉取代码的步骤将会消耗很长时间，如果对构建过程中拉取的代码进行持久化，那么将会大大减少代码下载的时间。

Jenkins 构建过程中会基于 Jenkins 的 workspace 目录进行一系列构建步骤，拉取的代码文件也将会放置于该目录中，在实际构建过程中需要对该目录进行持久化来避免每次代码都要全量拉取。在 Jenkinsfile 中，可以直接指定 PVC 的名称把对应的持久化卷挂载到 Pod 中的指定目录，这样需要提前创建 Jenkins 工作目录的持久化卷。本例持久化采用 NFS，首先创建对应的 PV，yaml 文件如下，请按需修改：

```yaml
apiVersion: v1
kind: PersistentVolume
metadata:
  name: jenkins-workspace
spec:
  capacity:
    storage: 200Gi
  accessModes:
    - ReadWriteMany
  volumeMode: Filesystem
  persistentVolumeReclaimPolicy: Recycle
  storageClassName: "jenkins-workspace"
  nfs:
    # real share directory
    path: /data/K8S/jenkins/
    # nfs real ip
    server: 192.168.2.2
```

创建对应的 PVC，yaml 文件如下：

```yaml
apiVersion: v1
kind: PersistentVolumeClaim
metadata:
  name: jenkins-workspace
  labels:
    app: cicd
spec:
  accessModes:
  - ReadWriteMany
  resources:
    requests:
      storage: 200Gi
  storageClassName: jenkins-workspace
```

创建完 PV 和 PVC 后，更改 Jenkinsfile 以便使用该 PVC 进行 Jenkins 工作目录的持久化，添加 persistentVolumeClaim 至 Jenkinsfile 的 volumes，之后再次构建就不会再重新拉取全部代码，Jenkinsfile 更改如下（添加最后一行至之前创建的 Jenkinsfile 中）：

```
volumes: [
  hostPathVolume(mountPath: '/var/run/docker.sock', hostPath: '/var/run/docker.sock'),
  hostPathVolume(mountPath: '/etc/hosts', hostPath: '/etc/hosts'),
  secretVolume(secretName: 'kube-config', mountPath: '/root/.kube/', key: 'config', path: 'config'),
  persistentVolumeClaim(claimName: 'jenkins-workspace', mountPath: '/home/jenkins/workspace/'),
```

完成上述配置以后，对 Jenkins 的工作目录就完成了持久化，下次再次构建时就不需要再重新拉取全部代码了。

4.10.2 解决 Maven 构建慢的问题

上一节为了加快代码的下载速度，对 Jenkins 的工作目录 workspace 进行了持久化，使其能够对代码文件进行存储，用于提高代码的下载速度。虽然代码文件进行了持久化，但是对 Java 代码进行编译的时候，Maven 首先会下载编译代码所用的依赖文件，这些文件并不是存储在 Jenkins 的工作目录下的。如果不对其进行文件持久化，那么每当使用 Maven 进行编译的时候，都会重新下载需要的依赖文件，这将会很明显地增加 Java 构建的时间，所以在实际使用时，对 Maven 依赖文件的存储目录进行持久化也是很重要的一步。

默认情况下，Maven 依赖文件的存储目录为用户目录下的.m2 文件夹（~/.m2），对其进行持久化有两种方式：

- 更改依赖文件目录为 Jenkins 的 workspace 目录。
- 对.m2 目录进行持久化。

因为在上一节已经实现了对 Jenkins 工作目录的持久化，可以直接将 Maven 依赖文件的目录改为 Jenkins workspace 目录下即可。只需要更改 Maven 配置文件 settings.xml 即可实现对此文件目录存储路径的更改，更改方式如下（在 settings.xml 文件中添加如下一行）：

```
<localRepository>/home/jenkins/workspace/mavenRepo</localRepository>
```

同样也可以对 Maven 默认文件存储目录进行持久化，持久化方式和对 Jenkins 的 workspace 的持久化配置方式相同，首先创建持久化所用的 PV，maven-repo.yaml 文件如下：

```
apiVersion: v1
kind: PersistentVolume
metadata:
  name: maven-repo
spec:
  capacity:
    storage: 200Gi
  accessModes:
    - ReadWriteMany
```

```
  volumeMode: Filesystem
  persistentVolumeReclaimPolicy: Recycle
  storageClassName: "maven-repo"
  nfs:
    # real share directory
    path: /data/K8S/maven/
    # nfs real ip
    server: 192.1686.2.2
```

之后创建对应的 PVC 文件：

```
apiVersion: v1
kind: PersistentVolumeClaim
metadata:
  name: maven-repo
  labels:
    app: cicd
spec:
  accessModes:
  - ReadWriteMany
  resources:
    requests:
      storage: 200Gi
  storageClassName: maven-repo
```

最后更改 Java 应用的 Jenkinsfile，添加最后一行至 Jenkinsfile 中即可：

```
volumes: [
    hostPathVolume(mountPath: '/var/run/docker.sock', hostPath: '/var/run/docker.sock'),
    hostPathVolume(mountPath: '/etc/hosts', hostPath: '/etc/hosts'),
    secretVolume(secretName: 'kube-config', mountPath: '/root/.kube/', key: 'config', path: 'config'),
    persistentVolumeClaim(claimName: 'jenkins-workspace', mountPath: '/home/jenkins/workspace/'),
    persistentVolumeClaim(claimName: 'maven-repo', mountPath: '~/.m2'),
```

通过上述两种方式的任意一种即可完成对 Maven 文件存储目录的持久化，之后再次编译 Java 应用时，会首先检测该目录是否已经存在所需要的依赖文件，如果已经存在这些文件就不会再次下载，这样会大大减少 Maven 的编译时间。

4.10.3 解决 NPM Install 的问题

通过上面的配置解决了代码拉取和 Maven 编译过慢的问题，但是对于 NodeJS 和 PHP 应用，都有可能需要进行依赖包的下载，比如 NodeJS 需要 npm install 进行依赖包的下载，PHP 应用需要 composer install 进行依赖包的下载，虽然 NodeJS 和 PHP 的应用依赖包存储在代码目录中（也就是 Jenkins 工作目录 workspace 对应的 NodeJS 或 PHP 项目的 Job 目录下），同时也配置了对该目录的持久化，但是对于 NodeJS 和 PHP 项目来说，虽然上次编译的时候已经下载了依赖的文件，再次执行 install 时，这个过程依旧很慢。对于 NodeJS 和 PHP 项目，这个 install 的过程并不是每次编译必须的过程，而之前的 Jenkinsfile 在构建步骤中，配置了每次编译之前都要进行 install，这将导致无须进行 install 的编译会浪费很多时间在这个过程中，可以通过判断决定是否需要进行 install 这个步骤来去除不必要的时间。

此时，需要更改 Jenkinsfile 文件，添加 if 判断语句。更改之前创建的 NodeJS 的 Jenkinsfile 文件如下（PHP 项目配置方式相同）：

```
stage('Exec NPM INSTALL or no...') {
    try {
      container('node6') {
        if ( env.NPM_INSTALL == 'true' ) {
           echo "Executing: npm install..."
           sh """
             npm install
           """
        } else {
          echo "no npm install"
        }
      }
    }
    catch (exc) {
      println "Failed to INSTALL - ${currentBuild.fullDisplayName}"
      throw(exc)
    }
}
stage('Run build') {
    try {
      container('node6') {
          sh """
             npm run build
          """
      }
    }
    catch (exc) {
      println "Failed to build - ${currentBuild.fullDisplayName}"
      throw(exc)
    }
}
```

在 stage(Exec NPM INSTALL or no...) 下添加判断语句。如果 NPM_INSTALL（变量在 Jenkins 中配置）变量为 true，将进行 npm install 操作；如果为 false，则不进行 npm install。

配置 Jenkins，添加变量 NPM_INSTALL，找到对应的 NodeJS 项目选择配置，然后找到参数化构建，依次单击"添加参数"选项➔"选择参数"，如图 4-30 所示。

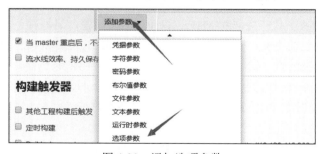

图 4-30　添加选项参数

填入如下信息，如图 4-31 所示。

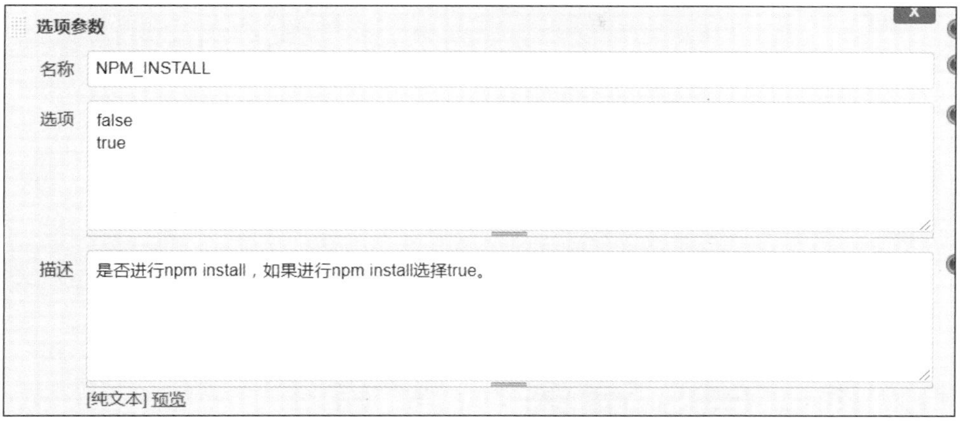

图 4-31　添加 NPM_INSTALL 参数

单击"保存"按钮后，选择 Build with Parameters，可以通过下拉框选择是否进行 npm install，如图 4-32 所示。

图 4-32　NPM_INSTALL 参数

4.11　小　结

本章主要演示了传统 Java 业务、Spring Cloud 系统组件、NodeJS 业务的自动化构建配置，演示了容器化业务部署的不同方式，比如 Deployment、StatefulSet、DaemonSet 等，均使用的是 Jenkins 新特性 Jenkins 流水线（Pipeline）进行持续集成和持续部署，在实际使用中不一定非要使用流水线进行构建，可以根据自己的业务场景选择其他风格的构建方式。使用流水线进行构建时，无论是 Java、NodeJS 还是 PHP 或者 Go，大致流程是类似的，主要是编译方式、基础镜像和 containerTemplate 不一样，可按照自己的业务场景和需求进行修改。对于 Jenkinsfile 也进行了优化处理，解决了实际使用中的很多问题，在真正使用的时候，为了使 Jenkinsfile 更加灵活和统一，可以对 Jenkinsfile 完全参数化，将 Jenkinsfile 所有可变的赋值都改为变量，然后在 Jenkins 上配置对应的变量即可，这样可以使 Jenkins 的构建变得更加灵活，也使得同一个 Jenkinsfile 适用于所有的项目。使用 Kubernetes 部署公司业务应用时，应遵循一次构建在任何地区、任何集群使用，部署过程中使用不同的环境变量或者是不同的启动参数来区分同一个镜像位于不同集群的配置。

第 5 章

Nginx Ingress 安装与配置

通常访问一个业务的流程为：

（1）用户在浏览器中输入域名。
（2）域名解析至业务的入口（一般为外部负载均衡器，比如阿里云的 SLB）。
（3）外部负载均衡器反向代理至 Kubernetes 的入口（一般为 Ingress）。
（4）通过 Ingress 再到对应的 Service 上。
（5）最后到达 Service 对应的某一个 Pod 上。

可见，在一般情况下，Ingress 主要是一个用于 Kubernetes 集群业务的入口。可以使用 Traefik、Istio、Nginx、HAProxy 作为 Ingress，因为相对于其他 Ingress，管理人员更熟悉 Nginx 或者 HAProxy 作为 Ingress，所以本章主要讲解 Nginx Ingress 安装与常用配置，有关 HAPproxy 的 Ingress，读者可参考相关资料。

5.1 Nginx Ingress 的安装

由于业务 Pod 一般部署在 Node 节点上，所以可以在每个 Node 上部署一个 Nginx Ingress，然后将域名解析至任意一个 Node 上即可访问业务 Pod。可以使用 DaemonSet 将 Nginx Ingress 部署至每个 Node 节点。

使用 DaemonSet 的方式部署 Nginx：

```
[root@K8S-master01 chap05]# kubectl create -f 5.1/mandatory.yaml
namespace/ingress-nginx created
configmap/nginx-configuration created
configmap/tcp-services created
configmap/udp-services created
serviceaccount/nginx-ingress-serviceaccount created
clusterrole.rbac.authorization.K8S.io/nginx-ingress-clusterrole created
```

```
    role.rbac.authorization.K8S.io/nginx-ingress-role created
    rolebinding.rbac.authorization.K8S.io/nginx-ingress-role-nisa-binding
created
    clusterrolebinding.rbac.authorization.K8S.io/nginx-ingress-clusterrole-nis
a-binding created
    daemonset.extensions/nginx-ingress-controller created
```

查看资源：

```
[root@K8S-master01 chap05]# kubectl get po -n ingress-nginx
NAME                              READY   STATUS    RESTARTS   AGE
nginx-ingress-controller-bzbzx    1/1     Running   0          5m
nginx-ingress-controller-gwkjb    1/1     Running   0          5m
nginx-ingress-controller-ltzrf    1/1     Running   0          5m
nginx-ingress-controller-tc6b6    1/1     Running   0          5m
nginx-ingress-controller-x5hvf    1/1     Running   0          5m
```

5.2　Nginx Ingress 的简单使用

本节我们来测试 Nginx Ingress 在集群中是否可以正常工作。

假如公司有一个 Web 服务的容器，可使用如下文件创建一个 Web 服务器：

```
[root@K8S-master01 5.2]# kubectl create -f web.yaml
deployment.extensions/nginx created
```

然后部署该 Web 容器的 Service：

```
[root@K8S-master01 5.2]# kubectl create -f web-service.yaml
service/nginx created
```

之后创建 Ingress 指向上面创建的 Service：

```
[root@K8S-master01 5.2]# cat web-ingress.yaml
apiVersion: extensions/v1beta1
kind: Ingress
metadata:
  name: nginx-ingress
spec:
  rules:
  - host: nginx.test.com
    http:
      paths:
      - backend:
          serviceName: nginx
          servicePort: 80
[root@K8S-master01 5.2]# kubectl create -f web-ingress.yaml
ingress.extensions/nginx-ingress created
```

创建的 Ingress 绑定的域名为 nginx.test.com，将域名解析至 Ingress 节点，通过域名 nginx.test.com 即可访问，如图 5-1 所示。

图 5-1　访问 nginx.test.com

假如公司要实现蓝绿发布，可以先部署新版本的 Web 容器，比如部署 v2 版本的 Web，访问该 Web，会返回 v2：

```
[root@K8S-master01 5.2]# kubectl create -f web-v2.yaml
deployment.extensions/nginx-v2 created
```

部署 Web v2 的 Service：

```
[root@K8S-master01 5.2]# kubectl create -f  web-service-v2.yaml
service/nginx-v2 created
```

更新 Ingress 指向 Webv2 的容器：

```
[root@K8S-master01 5.2]# kubectl apply -f web-ingress-v2.yaml
ingress.extensions/nginx-ingress configured
```

此时访问 nginx.test.com/foo/会返回 v2，如图 5-2 所示。

图 5.2　访问 nginx.test.comv2

5.3　Nginx Ingress Redirect

Redirect 主要用于域名的重定向，比如访问 a.com 被重定向到 b.com。
以下是访问 nginx.redirect.com 被重定向到 baidu.com 的示例：

```
[root@K8S-master01 5.4]# kubectl create -f redirect.yaml
ingress.extensions/redirect created
[root@K8S-master01 5.4]# cat redirect.yaml
apiVersion: extensions/v1beta1
kind: Ingress
```

```yaml
metadata:
  annotations:
    nginx.ingress.kubernetes.io/permanent-redirect: https://www.baidu.com
  name: redirect
  namespace: default
spec:
  rules:
  - host: nginx.redirect.com
    http:
      paths:
      - path: /
        backend:
          serviceName: nginx-v2
          servicePort: 80
```

使用 curl 访问域名 nginx.redirect.com，可以看到 301（请求被重定向的返回值）：

```
[root@K8S-master01 5.4]# curl -I nginx.redirect.com
HTTP/1.1 301 Moved Permanently
Server: nginx/1.15.8
Date: Sun, 24 Feb 2019 06:57:43 GMT
Content-Type: text/html
Content-Length: 169
Connection: keep-alive
Location: https://www.baidu.com
```

5.4 Nginx Ingress Rewrite

Rewrite 主要用于地址重写，比如访问 nginx.test.com/rewrite 跳转到 nginx.test.com，访问 nginx.test.com/rewrite/foo 会跳转到 nginx.test.com/foo 等。

这里我们根据 5.2 一节的配置，增加 Rewrite，示例如下：

```yaml
apiVersion: extensions/v1beta1
kind: Ingress
metadata:
  annotations:
    nginx.ingress.kubernetes.io/rewrite-target: /$1
  name: rewrite
  namespace: default
spec:
  rules:
  - host: nginx.test.com
    http:
      paths:
      - backend:
          serviceName: nginx-v2
          servicePort: 80
        path: /rewrite/?(.*)
```

5.5　Nginx Ingress 错误代码重定向

本节主要演示当访问链接返回值为 404、503 等错误时，如何自动跳转到自定义的错误页面。
定义 Service：

```
[root@K8S-master01 5.5]# kubectl create -f err-service.yaml
service/nginx-errors created
```

创建 Error Page Web Server（错误页面 Web 服务器）：

```
[root@K8S-master01 5.5]# kubectl create -f err-deploy.yaml
deployment.apps/nginx-errors created
```

配置 Nginx Ingress：

```
kubectl edit daemonset nginx-ingress-controller -n ingress-nginx
#添加
- --default-backend-service=$(POD_NAMESPACE)/nginx-errors
```

配置 ConfigMap：

```
kubectl edit cm nginx-configuration -n ingress-nginx
#添加如下：
data:
  apiVersion: v1
  client_max_body_size: 20m
  custom-http-errors: 404,415,503
```

更新 Nginx Ingress：

```
[root@K8S-master01 5.5]# kubectl patch daemonset nginx-ingress-controller -p
"{\"spec\":{\"template\":{\"metadata\":{\"annotations\":{\"date\":\"`date
+'%s'`\"}}}}}" -n ingress-nginx
daemonset.extensions/nginx-ingress-controller patched
```

更新完成以后访问一个不存在的页面，比如之前定义的 nginx.test.com，访问其不存在的页面 123，就会跳转到 Error Server 中的页面：

```
[root@K8S-master01 5.5]# curl http://nginx.test.com/123
<span>The page you're looking for could not be found.</span>
```

5.6　Nginx Ingress SSL

SSL 配置主要用于对域名的 HTTPS 访问，具体的配置步骤如下。
生成证书。如果是生产环境，证书为在第三方公司购买的证书：

```
[root@K8S-master01 5.6]# openssl req -x509 -nodes -days 365 -newkey rsa:2048
-keyout tls.key -out tls.crt -subj "/CN=www.xxx.com"
Generating a 2048 bit RSA private key
.....+++
```

```
.........................+++
writing new private key to 'tls.key'
-----
[root@K8S-master01 5.6]# kubectl create secret generic ca-secret
--from-file=tls.crt=tls.crt --from-file=tls.key=tls.key
secret/ca-secret created
```

创建 SSL：

```
[root@K8S-master01 5.6]# kubectl create -f test-ssl.yaml
ingress.extensions/ssl created
```

访问测试，会自动跳转到 https，如图 5-3 所示。

图 5-3　HTTPS 配置

5.7　Nginx Ingress 匹配请求头

请求头匹配主要用于匹配用户来源（比如使用 iPhone 或者 iPad 的用户和使用手机或者电脑的用户），然后将其访问转发到对应的业务上。

匹配 iPhone：

```
[root@K8S-master01 5.7]# kubectl create -f snippet.yaml
ingress.extensions/snippet created
[root@K8S-master01 5.7]# cat snippet.yaml
apiVersion: extensions/v1beta1
kind: Ingress
metadata:
  annotations:
    nginx.ingress.kubernetes.io/server-snippet: |
      set $agentflag 0;
      if ($http_user_agent ~* "iPhone" ){
        set $agentflag 1;
      }
      if ( $agentflag = 1 ) {
        return 301 http://nginx.test.com;
      }
  name: snippet
  namespace: default
spec:
  rules:
```

```
    - host: nginx.snippet.com
      http:
        paths:
        - path: /
          backend:
            serviceName: nginx-v2
            servicePort: 80
```

此时，访问 nginx.snippet.com 会访问 nginx-v2。请求头为 iPhone，会跳转到 nginx.test.com。直接访问测试，如图 5-4 所示。

图 5-4　直接访问域名

使用浏览器的开发者工具将终端类型改为 iPhone，或者直接用 iPhone 手机访问（线上业务一般配置的都有 DNS，可以直接解析域名，测试环境可能需要自己单独配置），如图 5-5 所示。

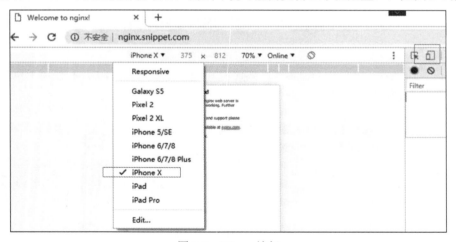

图 5-5　iPhone 访问

刷新页面会自动跳转，参考图 5-6。

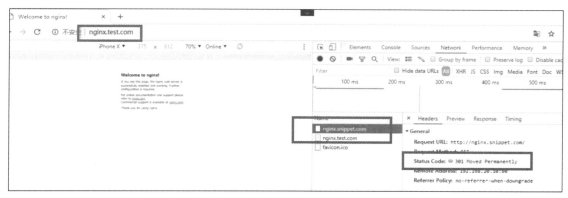

图 5-6　自动跳转

5.8　Nginx Ingress 基本认证

有些网站可能需要通过密码来访问，对于这类网站可以使用 Nginx 的 basic-auth 设置密码访问，具体方法如下。

创建密码：

```
[root@K8S-master01 5.8]# htpasswd -c auth foo
New password:
Re-type new password:
Adding password for user foo
[root@K8S-master01 5.8]# cat auth
foo:$apr1$okma2fx9$hdTJ.KFmi4pY9T6a2MjeS1
```

基于之前创建的密码创建 Secret：

```
[root@K8S-master01 5.8]# kubectl create secret generic basic-auth
--from-file=auth
    secret/basic-auth created
```

创建 Ingress：

```
[root@K8S-master01 5.8]# kubectl create -f test-auth.yaml
ingress.extensions/ingress-with-auth created
```

访问测试，如图 5-7 所示。

图 5-7　加密访问

5.9 Nginx Ingress 黑/白名单

有些网页可能只需要指定用户访问，比如公司的 ERP 只能在公司内部访问，此时可以使用白名单作为公司的出口 IP。有些网页不允许某些 IP 访问，比如一些有异常流量的 IP，此时，可以使用黑名单禁止该 IP 访问。

5.9.1 配置黑名单

配置黑名单禁止某一个或某一段 IP，需要在 Nginx Ingress 的 ConfigMap 中配置，具体方法如下。

配置 ConfigMap，将 192.168.10.129 添加至黑名单：

```
kubectl edit cm nginx-configuration -n ingress-nginx

apiVersion: v1
data:
  block-cidrs: 192.168.10.129
```

滚动更新 Nginx Ingress：

```
[root@K8S-master01 5.8]# kubectl patch daemonset nginx-ingress-controller -p "{\"spec\":{\"template\":{\"metadata\":{\"annotations\":{\"date\":\"`date +'%s'`\"}}}}}" -n ingress-nginx
daemonset.extensions/nginx-ingress-controller patched
```

再次访问，发现该 IP 已经被禁止，如图 5-8 所示。

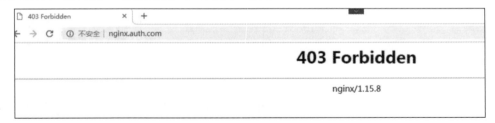

图 5-8　黑名单

5.9.2 配置白名单

白名单表示只允许某个 IP 可以访问，直接在 yaml 文件中配置即可，比如只允许 192.168.10.128 访问：

```
apiVersion: extensions/v1beta1
kind: Ingress
metadata:
  name: ingress-with-auth
  annotations:
```

```yaml
    nginx.ingress.kubernetes.io/whitelist-source-range: 192.168.10.129
    # type of authentication
    nginx.ingress.kubernetes.io/auth-type: basic
    # name of the secret that contains the user/password definitions
    nginx.ingress.kubernetes.io/auth-secret: basic-auth
    # message to display with an appropriate context why the authentication is required
    nginx.ingress.kubernetes.io/auth-realm: 'Authentication Required - foo'
spec:
  rules:
  - host: nginx.auth.com
    http:
      paths:
      - path: /
        backend:
          serviceName: nginx
          servicePort: 80
```

创建该 Ingress:

```
[root@K8S-master01 5.9]# kubectl apply -f test-auth.yaml
ingress.extensions/ingress-with-auth configured
```

192.168.10.129 访问是可以的, 如图 5-9 所示。

图 5-9 白名单访问

其他 IP 访问被禁止:

```
[root@K8S-master01 5.9]# curl nginx.auth.com
<html>
<head><title>403 Forbidden</title></head>
<body>
<center><h1>403 Forbidden</h1></center>
<hr><center>nginx/1.15.8</center>
</body>
</html>
```

5.10 Nginx Ingress 速率限制

有时候可能需要限制速率以降低后端压力, 此时, 可以使用 Nginx 的 ratelimit 进行配置, 具体方法如下。

首先没有加速率，使用 ab 进行访问，Failed 为 0：

```
[root@K8S-master01 5.9]# ab -c 10 -n 100 http://nginx.redirect.com/ | grep requests:
  Complete requests:    100
  Failed requests:      0
```

添加速率限制：

```
[root@K8S-master01 5.10]# kubectl apply -f test-rate-limit.yaml
ingress.extensions/redirect configured
```

再次使用 ab 测试，Failed 为 67：

```
[root@K8S-master01 5.10]# ab -c 10 -n 100 http://nginx.redirect.com/ | grep requests:
  Complete requests:    100
  Failed requests:      67
```

其他配置如下：

```
#限制每秒的连接，单个IP:
nginx.ingress.kubernetes.io/limit-rps:

#限制每分钟的连接，单个IP:
nginx.ingress.kubernetes.io/limit-rpm

#限制客户端每秒传输的字节数 单位为K:
nginx.ingress.kubernetes.io/limit-rate
```

5.11 使用 Nginx 实现灰度/金丝雀发布

本节演示利用 Nginx 简单实现灰度发布。

5.11.1 创建 v1 版本

首先创建模拟 Production（生产）环境的 Namespace：

```
[root@K8S-master01 5.11]# kubectl create ns canary-production
namespace/canary-production created
```

创建 v1 版本的服务：

```
[root@K8S-master01 5.11]# kubectl create -f canary-v1.yaml
deployment.extensions/canary-v1 created
```

创建 v1 版本的 Service：

```
[root@K8S-master01 5.11]# kubectl create -f canary-v1-service.yaml
service/canary-v1 created
```

5.11.2 创建 v2 版本

创建 v2 版本的 Namespace：

```
[root@K8S-master01 5.11]# kubectl create ns canary-production-canary
namespace/canary-production-canary created
```

创建 v2 版本的服务：

```
[root@K8S-master01 5.11]# kubectl create -f canary-v2.yaml
deployment.extensions/canary-v2 created
```

创建 v2 版本的 Service：

```
[root@K8S-master01 5.11]# kubectl create -f canary-v2-service.yaml
service/canary-v2 created
```

5.11.3 创建 Ingress

创建 v1 版本的 Ingress：

```
[root@K8S-master01 5.11]# kubectl create -f canary-v1-ingress.yaml
ingress.extensions/canary-v1 created
```

访问测试：

```
[root@K8S-master01 5.11]# curl canary.com
<h1>Canary v1</h1>
```

此时只能访问至 v1 版本。

创建 v2 版本的 Ingress：

```
[root@K8S-master01 5.11]# cat canary-v2-ingress.yaml
apiVersion: extensions/v1beta1
kind: Ingress
metadata:
  name: canary-v2
  annotations:
    kubernetes.io/ingress.class: nginx
    nginx.ingress.kubernetes.io/canary: "true"
    nginx.ingress.kubernetes.io/canary-weight: "10"
spec:
  rules:
  - host: canary.com
    http:
      paths:
      - backend:
          serviceName: canary-v2
          servicePort: 8080
```

此时通过 nginx.ingress.kubernetes.io/canary-weight: "10" 设置的权重是 10，即 v1:v2 为 9:1。

5.11.4　测试灰度发布

使用 Ruby 脚本进行测试，此脚本会输出 v1 和 v2 的访问次数：

```
[root@K8S-master01 5.11]# cat count.rb
counts = Hash.new(0)

100.times do
  output = `curl -s canary.com | grep 'Canary' | awk '{print $2}' | awk -F"<" '{print $1}'`
  counts[output.strip.split.last] += 1
end

puts counts
[root@K8S-master01 5.11]# ruby count.rb
{"v1"=>84, "v2"=>16}
[root@K8S-master01 5.11]# ruby count.rb
{"v1"=>92, "v2"=>8}
[root@K8S-master01 5.11]# ruby count.rb
{"v1"=>91, "v2"=>9}
```

可以看到比例差不多是 1:9。

更改权重，将 v2 权重改为 20：

```
[root@K8S-master01 5.11]# kubectl apply -f canary-v2-ingress-20.yaml
ingress.extensions/canary-v2 configured
```

再次测试：

```
[root@K8S-master01 5.11]# ruby count.rb
{"v1"=>81, "v2"=>19}
[root@K8S-master01 5.11]# ruby count.rb
{"v1"=>84, "v2"=>16}
[root@K8S-master01 5.11]# ruby count.rb
{"v1"=>84, "v2"=>16}
```

比例差不多为 8:2。

以上演示了简单的灰度发布，根据业务需求匹配请求头等内容，读者在实际使用过程中可以根据自己的业务需求进行配置。

5.12　小　结

本章介绍使用了 Nginx Ingress 的配置规则，Ingress 作为集群的入口，可以在 Ingress 上进行简单的流量、权限控制等。本章使用的 Nginx Ingress 是 Kubernetes 社区维护的版本，还有一个是 Nginx 社区维护的 Nginx Ingress，两者的区别可以查看：https://github.com/nginxinc/kubernetes-ingress/blob/master/docs/nginx-ingress-controllers.md。与 Nginx Ingress 相对应的还有 HAProxy、Istio、Traefik 等，其他的可以查看：https://kubernetes.io/docs/concepts/services-networking/ingress-controllers/，读者可以选择其一进行使用，当然也可以选择最为常用的 Nginx 或者 HAProxy。

第 6 章

Server Mesh 服务网格

Server Mesh 是由 Buoyant 公司的 CEO William Morgan 发起的,它对 Server Mesh 的定义如下。

- 专用基础设施层:独立的运行单元。
- 包括数据层和控制层:数据层负责交付应用请求,控制层控制服务如何通信。
- 轻量级透明代理:实现形式为轻量级网络代理。
- 处理服务间通信:主要目的是实现复杂网络中服务间通信。
- 可靠地交付服务请求:提供网络弹性机制,确保可靠交付请求。
- 与服务部署一起,但服务无须感知:尽管跟应用部署在一起,但对应用是透明的。

本章主要介绍 Server Mesh 的基本概念,当前流行的开源服务网格 Istio 的架构、安装、配置和基本使用。

6.1 服务网格的基本概念

Service Mesh 作为透明代理可以运行在任何基础设施环境,而且和应用非常靠近,Server Mesh 的功能大致如下:

- 负载均衡。运行环境中微服务实例通常处于动态变化状态,而且经常可能出现个别实例不能正常提供服务、处理能力减弱、卡顿等现象。但由于所有请求对 Service Mesh 来说是可见的,因此可以通过提供高级负载均衡算法来实现更加智能、高效的流量分发,降低延时,提高可靠性。
- 服务发现。以微服务模式运行的应用变更非常频繁,应用实例的频繁增加和减少带来的问题是,如何精确地发现新增实例以及避免将请求发送给已不存在的实例而变得更加复杂。Service Mesh 可以提供简单、统一、平台无关的多种服务发现机制,如基于 DNS、键-值对(key-value pair)存储的服务发现机制。

- 熔断。动态环境中服务实例中断或不健康导致的服务中断可能会经常发生，这就要求应用或者其他工具具有快速监测并从负载均衡池中移除不提供服务实例的能力，这种能力也称熔断，以此使得应用无须消耗更多不必要的资源不断地尝试，而是快速失败或者降级，从而避免一些潜在的关联性错误，而 Service Mesh 可以很容易实现基于请求和连接级别的熔断机制。
- 动态路由。随着服务提供商以提供高稳定性、高可用性及高 SLA 服务为主要目标，出现的各种应用部署策略都尽可能地达到无服务中断部署，以此避免变更而导致服务的中断和稳定性降低，例如 Blue/Green 部署、Canary 部署，但是实现这些高级部署策略通常非常困难。关于应用部署策略可参考 https://thenewstack.io/deployment-strategies/的内容，其对各种部署策略做了详细的比较。如果运维人员想要轻松地将应用流量从 staging 环境切换到生产环境，从一个版本到另外一个版本，或者从一个数据中心到另外一个数据中心，甚至可以通过一个中心控制层控制多少比例的流量被切换。那么 Service Mesh 提供的动态路由机制和特定的部署策略（如 Blue/Green 部署）结合起来，实现上述目标将会变得更加容易。
- 安全通信。无论何时，安全在整个公司、业务系统中都占据着举足轻重的位置，也是非常难以实现和控制的部分。而微服务环境中，不同的服务实例间通信变得更加复杂，那么如何保证这些通信是在安全、授权情况下进行就非常重要。通过将安全机制如 TLS 加解密和授权实现在 Service Mesh 上，不仅可以避免在不同应用上的重复实现，而且很容易在整个基础设施层更新安全机制，甚至无须对应用做任何操作。
- 多语言支持。由于 Service Mesh 作为独立运行的透明代理，很容易支持多语言。
- 多协议支持。同多语言支持一样，实现多协议支持也非常容易。
- 指标和分布式追踪。Service Mesh 对整个基础设施层的可见性使得它不仅可以暴露单个服务的运行指标，而且可以暴露整个集群的运行指标。
- 重试和最后期限。Service Mesh 的重试功能可避免将其嵌入到业务代码中，同时最后期限使得应用允许一个请求的最长生命周期，而不是无休止地重试。

6.2　服务网格产品

常见的服务网格产品如下：

- Linkerd: Buoyant 公司在 2016 年率先开源的高性能网络代理程序，它的出现标志着 Server Mesh 时代的开始。
- Envoy: 同 Linkerd 一样，Envoy 也是一款高性能的网络代理程序，为云原生应用而设计。
- Istio: Istio 受 Google、IBM、Lyft 及 RadHat 等公司的大力支持和推广，于 2017 年 5 月发布，底层为 Envoy。
- Conduit: 2017 年 12 月发布，是 Buoyant 公司的第二款 Server Mesh 产品，根据 Linkerd 在生产线上的实际使用经验而设计，并以最小复杂性作为设计基础。

> **注 意**
>
> 6.1 节和 6.2 节摘自《Server Mesh 实战：基于 Linkerd 和 Kubernetes 的微服务实践》一书。

6.3 Istio 介绍

Istio 是一个完全开源的服务网格，可以透明地分层到现有的分布式应用程序上。Istio 有助于降低应用部署的复杂性，并减轻开发团队的压力。Istio 的多样化能够高效地运行于分布式微服务架构中，并提供保护、链接和监控服务的统一方法。

Istio 提供一种简单的方式来为已部署的服务建立网络，该网络具有负载均衡、服务间认证、监控等功能。

请参考 Istio 中文网址：https://istio.io/zh/docs/concepts/what-is-istio/

6.3.1 Istio 架构

Istio 服务网格逻辑上分为数据平面和控制平面两部分：

- 数据平面由一组 sidecar 方式部署的智能代理（Envoy）组成，这些代理可以调节和控制微服务及 Mixer 之间所有的网络通信。
- 控制平面负责管理和配置代理来路由流量，此外控制平面配置 Mixer 以实施策略和收集遥测数据。

Istio 的架构图，如图 6-1 所示。

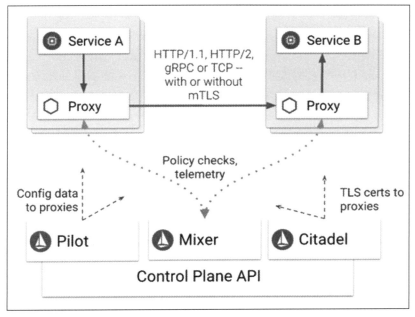

图 6-1　Istio 架构图

6.3.2 名词解释

1. Envoy

Istio 使用 Envoy 代理的扩展版本，Envoy 是以 C++开发的高性能代理，用于调解服务网格中所有服务的所有入站和出站流量。Envoy 内置功能如下：

- 动态服务发现
- 负载均衡
- TLS 终止
- HTTP/2 & gRPC 代理
- 熔断器
- 健康检查、基于百分比流量拆分的灰度发布
- 故障注入
- 丰富的度量指标

Envoy 被部署为 sidecar，和对应服务在同一个 Kubernetes Pod 中。这允许 Istio 将大量关于流量行为的信号作为属性提取出来，而这些属性又可以在 Mixer 中用于执行策略决策并发送给监控系统，以提供整个网络行为的信息。

Sidecar 代理模型还可以将 Istio 的功能添加到现有部署中，而无须重新构建或重写代码。

2. Mixer

一个独立于平台的组件，负责在服务网格上执行访问控制和使用策略，并从 Envoy 代理和其他服务收集遥测数据。Mixer 中包括一个灵活的插件模型，使其能够接入到各种主机环境和基础设置后端，从这些细节中抽象出 Envoy 代理和 Istio 管理的服务。

3. Pilot

为 Envoy sidecar 提供服务发现的功能，为智能路由（例如 A/B 测试、金丝雀部署等）和弹性（超时、重试、熔断器等）提供流量管理功能。它将控制流量行为的高级路由规则转换为特定于 Envoy 的配置，并在运行时将它们传播到 sidecar。Pilot 将平台特定的服务发现机制抽象化，并将其合成为符合 Envoy 数据平面 API 的任何 sidecar 都可以使用的标准格式，这种松耦合使得 Istio 能够在多种环境下运行（如 kubernetes、consul、nomad），同时保持用于流量管理的相同操作界面。

4. Citadel

通过内置身份和凭证管理可以提供强大的服务与服务之间的最终用户身份验证，可用于升级服务网格中未加密的流量，并为运维人员提供基于服务标识而不是网络控制的强制执行策略的能力。从 0.5 版本开始，Istio 支持基于角色的访问控制，以控制服务访问。

6.3.3 流量管理

Istio 流量管理的核心组件是 Pilot，该软件管理和配置部署在特定 Istio 服务网格中的所有 Envoy 代理实例，指定 Envoy 代理之间使用什么样的路由流量规则，并配置故障恢复功能，如超时、重

试、熔断器等，还维护网格中所有服务的规范模型，并使用这个模型通过发现服务让 Envoy 了解网格中的其他实例。

每个 Envoy 实例都会维护从 Pilot 获得的负载均衡信息以及其负载均衡池中的其他实例的定期健康检查，从而允许其在目标实例之间智能分配流量，同时遵循其指定的路由规则。

1. 流量管理的优点

使用 Istio 的流量管理模型，本质上是将流量与基础设施扩容解耦，让运维人员可以通过 Pilot 指定流量遵循什么规则，而不是指定哪些 Pod/VM 应该接收流量，关于这一点，Pilot 和智能 Envoy 代理会帮我们搞定。例如，可以通过 Pilot 指定特定服务的 5%流量转到金丝雀版本，而不必考虑金丝雀部署的大小，或根据请求的内容将流量发送到特定版本，如图 6-2 所示。

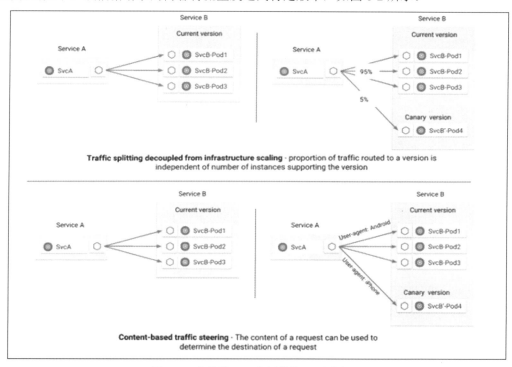

图 6-2　流量管理（此图模糊，不清晰）

将流量从基础设施扩展中解耦，这样就可以让 Istio 提供各种流量的管理功能，这些功能在应用程序代码之外。除了 A/B 测试的动态请求路由逐步推出和金丝雀发布之外，它还使用超时、重试和熔断器处理故障恢复，最后还可以通过故障注入来测试服务之间故障恢复策略的兼容性，这些功能都是通过在服务网格中部署的 Envoy sidecar 代理来实现的。

Pilot 公开了用于服务发现、负载均衡池和路由表动态更新的 API，这些 API 将 Envoy 从平台特有的细微差别中解脱出来，简化了设计并提升了跨平台的可移植性。

运维人员可以通过 Pilot 的 Rules API 指定高级流量管理规则，这些规则被翻译成低级配置并通过 discovery API 分发到 Envoy 实例。

2. 请求路由

如 Pilot 所述,特定网格中服务的规范表示由 Pilot 维护,服务的 Istio 模型和在底层平台(K8S、mesos 以及 cloudfoundry)中的表达无关。特定平台的适配器负责从各自平台中获取元数据的各种字段,然后对服务模型进行填充。

Istio 引入了服务版本的概念,可以通过版本(v1、v2)或环境(staging、prod)对服务进行进一步的细分。使用这种方式的常见场景包括 A/B 测试或金丝雀部署。Istio 的流量路由规则可以根据服务版本来对服务之间的流量进行附加控制。

3. 服务间通信

服务间通信如图 6-3 所示。

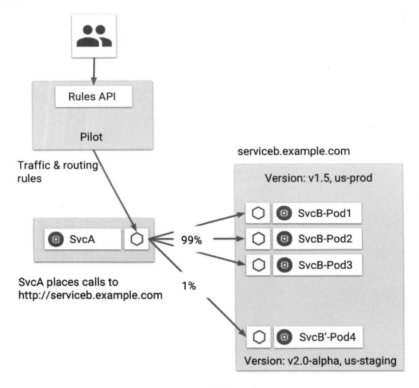

图 6-3 服务间通信

服务的客户端不知道服务不同版本间的差异,它们可以使用服务的主机名或者 IP 地址继续访问服务。Envoy sidecar 代理负责拦截并转发客户端和服务器之间的所有请求和响应。

运维人员使用 Pilot 指定路由规则,Envoy 根据这些规则动态地确定其服务版本的实际选择。该模型使应用程序代码能够将它从其依赖服务的演进中解耦出来,同时提供其他好处。路由规则让 Envoy 能够根据诸如 Header、与源目的地相关的标签或分配给每个版本的权重等标准来进行版本选择。

Istio 还为同一服务版本的多个实例提供流量负载均衡。

Istio 不提供 DNS,可以使用 coredns 和 kube-dns 来解析 FQDN。

4. 服务发现和注册

（1）服务注册

Istio 假定存在服务注册表，以跟踪应用程序中服务的 Pod/VM。它还假定服务的新实例自动注册到服务注册表，并且自动删除不健康的实例，诸如 K8S、mesos 等平台已经为基于容器的应用程序提供了这样的功能，为基于虚拟机的应用程序提供的解决方案就更多了。

（2）服务发现

Pilot 使用来自服务注册的信息并提供与平台无关的服务发现接口，网格中的 Envoy 实例执行服务发现，并相应地动态更新其负载均衡池，如图 6-4 所示。

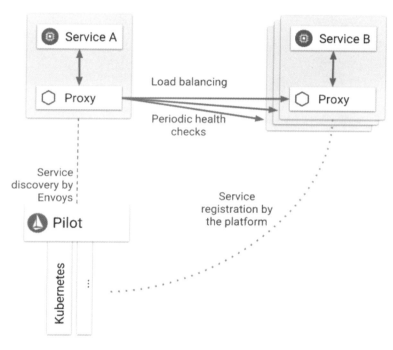

图 6-4 服务发现

图 6-4 中，网格中的服务使用 DNS 名称访问彼此，服务的所有 HTTP 流量都会通过 Envoy 自动重新路由，Envoy 在负载均衡池中的实例之间分发流量。虽然 Envoy 支持多种负载均衡算法，但是 Istio 目前仅允许 3 种负载平衡模式：轮询、随机和带权重的最少请求。

除了负载均衡外，Envoy 会定期检查池中的每个实例，根据监控检查 API 调用的失败率将实例分类为不健康或健康，从而进行弹出或重新添加回负载均衡池。

5. 故障注入

虽然 Envoysidecar/Proxy 为在 Istio 上运行的服务提供了大量的故障恢复机制，但测试整个应用程序到端的故障恢复能力依然是必须的。错误配置的故障恢复策略（例如，跨服务调用的不兼容/限制性超时）可能导致应用程序中的关键服务持续不可用，从而破坏用户体验。

Istio 能在不杀死 Pod 的情况下将协议特定的故障注入网络中，在 TCP 层制造数据包的延迟或损坏。无论网络级别的故障如何，应用层观察到的故障都是一样的，并且可以在应用层注入更有意义的故障，例如 HTTP 错误代码，以检验和改善应用的弹性。

运维人员可以为符合特定条件的请求配置故障，还可以进一步限制遭受故障的请求百分比。可以注入两种类型的故障：延迟和中断。延迟是计时故障，模拟网络延迟上升或上游服务超载的情况。中断是模拟上游服务的崩溃故障，中断通常以 HTTP 错误代码或 TCP 连接失败的形式表现。

6.4　Istio 的安装

官方建议使用 Helm 安装 Istio，本节我们就介绍这种安装方法。

6.4.1　安装文件下载

自行安装时，将版本改为最新版：

```
wget https://github.com/istio/istio/releases/download/1.0.4/istio-1.0.4-linux.tar.gz
```

复制 Istio 工具至 /usr/local/bin：

```
[root@K8S-master01 istio-1.0.4]# cp bin/istioctl /usr/local/bin/
[root@K8S-master01 istio-1.0.4]# istioctl version
Version: 1.0.4
GitRevision: d5cb99f479ad9da88eebb8bb3637b17c323bc50b
User: root@8c2feba0b568
Hub: docker.io/istio
GolangVersion: go1.10.4
BuildStatus: Clean
```

6.4.2　安装 Istio

使用 Helm 安装 Istio，可以更改 values 文件，选择性安装：

```
helm install install/kubernetes/helm/istio --name istio --namespace istio-system
```

> **注　意**
>
> 如果选用 Istio 代替 Ingress，需要将之前创建的 Nginx 和 traefik 卸载掉。

查看资源：

```
[root@K8S-master01 istio-1.0.4]# kubectl get po -n istio-system
NAME                                      READY   STATUS    RESTARTS   AGE
istio-citadel-5c9544c886-6vdvn            1/1     Running   0          43m
istio-egressgateway-b8677f5bc-28v6p       1/1     Running   0          43m
istio-galley-8dcbb5f99-w8gg8              1/1     Running   0          43m
istio-ingressgateway-8488676c6b-hx4cq     1/1     Running   0          43m
istio-pilot-987746df9-74gfb               2/2     Running   0          43m
```

```
istio-policy-79657bf7ff-rkxjp               2/2    Running    0    43m
istio-security-post-install-9x55f           0/1    Completed  0    43m
istio-sidecar-injector-6bd4d9487c-7rlmq     1/1    Running    0    43m
istio-telemetry-5dc97c4858-cvssz            2/2    Running    0    43m
prometheus-65d6f6b6c-kbvz7                  1/1    Running    0    43m
```

6.4.3 配置自动注入 sidecar

修改 APIServer 允许 Istio 自动注入：

```
vi /etc/kubernetes/manifests/kube-apiserver.yaml
-
--enable-admission-plugins=NamespaceLifecycle,LimitRanger,ServiceAccount,DefaultStorageClass,DefaultTolerationSeconds,MutatingAdmissionWebhook,ValidatingAdmissionWebhook,ResourceQuota
```

测试自动注入。创建测试 Namespace，并进行 label：

```
kubectl create ns istio-test
kubectl label namespace istio-test istio-injection=enabled
```

创建测试应用，此时默认创建的 Pod 有两个容器，一个是 sleep，一个是 sidecar：istio-proxy：

```
[root@K8S-master01 istio-1.0.4]# kubectl apply -f samples/sleep/sleep.yaml -n istio-test
service/sleep created
deployment.extensions/sleep created
```

查看部署的容器：

```
[root@K8S-master01 istio-1.0.4]# kubectl get po -n istio-test
NAME                  READY   STATUS    RESTARTS   AGE
sleep-86cf99dfd6-h2nzh  2/2   Running   0          92s
```

6.5 Istio 配置请求路由

本示例采用官方的 Bookinfo 进行测试，它由 4 个单独的微服务构成，用来演示多种 Istio 的特性。这个应用模仿在线书店的一个分类，显示一本书的信息。页面上会显示一本书的描述，图书的细节（ISBN、页数等）以及这本书的一些评论。

Bookinfo 应用分为 4 个单独的微服务：

- productpage：productpage 微服务会调用 details 和 reviews 两个微服务，用来生成页面。
- details：这个微服务包含了图书的信息。
- reviews：这个微服务包含了图书相关的评论，它还会调用 ratings 微服务。
- ratings：ratings 微服务中包含了由图书评价组成的评级信息。
  其中，reviews 微服务有 3 个版本：
  ➢ v1 版本，不会调用 ratings 服务。
  ➢ v2 版本，会调用 ratings 服务，并使用 1 到 5 个黑色星形图标来显示评分信息。

> v3 版本，会调用 ratings 服务，并使用 1 到 5 个红色星形图标来显示评分信息。

部署测试用例：

Istio 提供了用于测试功能的应用程序 bookinfo，可以直接通过下述命令创建 bookinfo：

```
[root@K8S-master01 istio-1.0.4]# kubectl apply -f
samples/bookinfo/platform/kube/bookinfo.yaml -n istio-test
service/details created
deployment.extensions/details-v1 created
service/ratings created
deployment.extensions/ratings-v1 created
service/reviews created
deployment.extensions/reviews-v1 created
deployment.extensions/reviews-v2 created
deployment.extensions/reviews-v3 created
service/productpage created
deployment.extensions/productpage-v1 created
```

将 Istio 的 Ingress gateway 的 Service type 改成 NodePort，默认为 LoadBalancer，然后使用 apply 部署：

```
kubectl apply -f samples/bookinfo/networking/bookinfo-gateway.yaml -n istio-test
```

创建 Istio 路由：

```
[root@K8S-master01 istio-1.0.4]# istioctl create -f
samples/bookinfo/networking/destination-rule-all.yaml -n istio-test
Created config destination-rule/istio-test/productpage at revision 1126754
Created config destination-rule/istio-test/reviews at revision 1126755
Created config destination-rule/istio-test/ratings at revision 1126756
Created config destination-rule/istio-test/details at revision 1126757
```

查看路由。此时浏览器可以通过 NodeIP:NodePort/productpage 访问，访问轮询的 v1、v2、v3：

```
[root@K8S-master01 istio-1.0.4]# istioctl get destinationrules -n istio-test
DESTINATION-RULE NAME    HOST            SUBSETS                     NAMESPACE      AGE
details                  details         v1,v2                       istio-test     52s
productpage              productpage     v1                          istio-test     52s
ratings                  ratings         v1,v2,v2-mysql,v2-mysql-vm  istio-test     52s
reviews                  reviews         v1,v2,v3                    istio-test     52s
```

设置只能访问 v1。此时浏览器可以通过 NodeIP:NodePort/productpage 访问，但是只能访问到 v1 版本：

```
[root@K8S-master01 istio-1.0.4]# istioctl create -f
samples/bookinfo/networking/virtual-service-all-v1.yaml -n istio-test
Created config virtual-service/istio-test/productpage at revision 1138043
Created config virtual-service/istio-test/reviews at revision 1138044
Created config virtual-service/istio-test/ratings at revision 1138045
Created config virtual-service/istio-test/details at revision 1138046
```

匹配指定用户访问 v2 可以看到星级，此时浏览器可以通过 NodeIP:NodePort/productpage 访问，只要登录用户为 virtual-service-reviews-test-v2.yaml 文件指定的用户就能访问到 v2 版本：

```
[root@K8S-master01 istio-1.0.4]# istioctl replace -f samples/bookinfo/networking/virtual-service-reviews-test-v2.yaml -n istio-test
Updated config virtual-service/istio-test/reviews to revision 1138197
```

6.6　Istio 熔断

断路器是创建弹性微服务应用程序的重要模式，断路器允许编写限制故障、延迟峰值以及其他不良网络特性使应用程序发出异常，一般用于测试程序的异常处理机制是否合理。本节熔断采用官方提供的 httpbin 进行测试。

6.6.1　创建测试用例

创建 httpbin 测试用例：

```
[root@K8S-master01 istio-1.0.4]# kubectl apply -f samples/httpbin/httpbin.yaml -n istio-test
service/httpbin created
deployment.extensions/httpbin created
```

6.6.2　配置熔断规则

创建一个目标规则，需针对 httpbin 服务设置断路器：

```
cat <<EOF | istioctl -n istio-test create -f -
apiVersion: networking.istio.io/v1alpha3
kind: DestinationRule
metadata:
  name: httpbin
spec:
  host: httpbin
  trafficPolicy:
    connectionPool:
      tcp:
        maxConnections: 1
      http:
        http1MaxPendingRequests: 1
        maxRequestsPerConnection: 1
    outlierDetection:
      consecutiveErrors: 1
      interval: 1s
      baseEjectionTime: 3m
      maxEjectionPercent: 100
EOF
```

说明：

maxConnections:1 及 http1MaxPendingRequests:1 表示如果超过了一个连接同时发起请求，Istio 就会熔断，阻止后续的请求或连接。

6.6.3 测试熔断

（1）部署测试工具 fortio，用于对容器业务进行压力测试：

```
[root@K8S-master01 istio-1.0.4]# kubectl -n istio-test apply -f samples/httpbin/sample-client/fortio-deploy.yaml
  deployment.apps/fortio-deploy created
```

（2）获取 fortio 容器 id：

```
[root@K8S-master01 istio-1.0.4]# FORTIO_POD=$(kubectl get pod -n istio-test | grep fortio | awk '{ print $1 }')
[root@K8S-master01 istio-1.0.4]# echo $FORTIO_POD
fortio-deploy-75c4fbd7f9-5qp9j
```

（3）发送一个请求，可以看到当前状态码为 200，即连接成功：

```
[root@K8S-master01 istio-1.0.4]# kubectl -n istio-test exec -it $FORTIO_POD -c fortio /usr/local/bin/fortio -- load -curl http://httpbin:8000/get
HTTP/1.1 200 OK
server: envoy
date: Fri, 11 Jan 2019 09:29:43 GMT
content-type: application/json
access-control-allow-origin: *
access-control-allow-credentials: true
content-length: 365
x-envoy-upstream-service-time: 19

{
  "args": {},
  "headers": {
    "Content-Length": "0",
    "Host": "httpbin:8000",
    "User-Agent": "istio/fortio-1.0.1",
    "X-B3-Sampled": "0",
    "X-B3-Spanid": "3e4b565b21cc553d",
    "X-B3-Traceid": "3e4b565b21cc553d",
    "X-Request-Id": "faadddc9-b5ec-4429-92aa-1b9b56a36dac"
  },
  "origin": "127.0.0.1",
  "url": "http://httpbin:8000/get"
}
```

（4）触发熔断。两个并发连接（-c 2），发送 20 请求（-n 20）：

```
[root@K8S-master01 istio-1.0.4]# kubectl -n istio-test exec -it $FORTIO_POD -c fortio /usr/local/bin/fortio -- load -c 2 -qps 0 -n 20 -loglevel Warning http://httpbin:8000/get
```

可以看到 503 的结果为 1，说明触发了熔断。

```
....
Code 200 : 19 (95.0 %)
Code 503 : 1 (5.0 %)
....
```

（5）提高并发量，可以看到故障率更高：

```
kubectl -n istio-test exec -it $FORTIO_POD -c fortio /usr/local/bin/fortio --load -c 5 -qps 0 -n 20 -loglevel Warning http://httpbin:8000/get
.....
Code 200 : 8 (40.0 %)
Code 503 : 12 (60.0 %)
.....
```

（6）查看 fortio 请求记录：

```
[root@K8S-master01 istio-1.0.4]# kubectl -n istio-test exec -it $FORTIO_POD -c istio-proxy -- sh -c 'curl localhost:15000/stats' | grep httpbin | grep pending
    cluster.outbound|8000||httpbin.istio-test.svc.cluster.local.upstream_rq_pending_active: 0
    cluster.outbound|8000||httpbin.istio-test.svc.cluster.local.upstream_rq_pending_failure_eject: 0
    cluster.outbound|8000||httpbin.istio-test.svc.cluster.local.upstream_rq_pending_overflow: 13
    cluster.outbound|8000||httpbin.istio-test.svc.cluster.local.upstream_rq_pending_total: 109
```

说明：

upstream_rq_pending_overflow 表示熔断的次数。

6.7　Istio 故障注入

有时我们可能需要通过一些故障来检查代码的逻辑性，这时可以使用 Istio 的故障注入来测试代码的错误处理。

6.7.1　基于 HTTP 延迟触发故障

在测试应用程序时，可以通过延迟 HTTP 的响应使后端程序触发超时异常，然后可以测试后端的错误处理方式是否设计的合理。

1. 配置测试用例

假如有一个 Web 项目，v1 采用 Nginx，v2 采用 httpd，默认流量走 Nginx 的 v1，当访问 URL 为/，且 header 的 end-user 为 jason 时访问 v2。

（1）部署该项目。

```
[root@K8S-master01 samples]# cat ~/nginx.yaml
```

```yaml
apiVersion: v1
kind: Service
metadata:
  name: nginx-svc
spec:
  template:
    metadata:
      labels:
        name: nginx-svc
spec:
  selector:
    run: ngx-pod
  ports:
  - protocol: TCP
    port: 80
    targetPort: 80
---
apiVersion: apps/v1beta1
kind: Deployment
metadata:
  name: ngx-pod
spec:
  replicas: 2
  template:
    metadata:
      labels:
        run: ngx-pod
        version: v1
    spec:
      containers:
      - name: nginx
        image: nginx:1.10
        ports:
        - containerPort: 80
```

```
[root@K8S-master01 samples]# cat ~/apache.yaml
```
```yaml
apiVersion: apps/v1beta1
kind: Deployment
metadata:
  name: ngx-pod-v2
spec:
  replicas: 1
  template:
    metadata:
      labels:
        run: ngx-pod
        version: v2
    spec:
      containers:
      - name: http
        image: httpd
        ports:
        - containerPort: 80
```

> **注　意**
>
> 此时 httpd 和 Nginx 使用的是同一个 Service 进行访问，使用 kubectl 创建以上文件即可。

（2）Istio 配置 gateway。添加域名 nginx.xxx.net，并配置 VirtualService 将访问/时流量指向 v1，当匹配用户是 jason 时将流量指向 v2：

```yaml
[root@K8S-master01 6.7.1.1]# cat ge.yaml
apiVersion: networking.istio.io/v1alpha3
kind: Gateway
metadata:
  name: nginx-gateway
spec:
  selector:
    istio: ingressgateway # use istio default controller
  servers:
  - port:
      number: 80
      name: http
      protocol: HTTP
    hosts:
    - "nginx.xxx.net"
    #- "*"
[root@K8S-master01 6.7.1.1]# cat vs.yaml
apiVersion: networking.istio.io/v1alpha3
kind: VirtualService
metadata:
  name: nginx
spec:
  hosts:
  - "nginx.xxx.net"
  gateways:
  - nginx-gateway
#  - mesh
  http:
  - match:
    - uri:
        exact: /
      headers:
        end-user:
          exact: jason
    route:
    - destination:
        host: nginx-svc # name of app service
        subset: v2
  - route:
    - destination:
        host: nginx-svc # name of app service
        subset: v1
        #port:
        #  number: 80
[root@K8S-master01 6.7.1.1]# cat dr.yaml
apiVersion: networking.istio.io/v1alpha3
kind: DestinationRule
```

```
metadata:
  name: nginx
spec:
  host: nginx-svc
  trafficPolicy:
    loadBalancer:
      simple: RANDOM
  subsets:
  - name: v1
    labels:
      version: v1
  - name: v2
    labels:
      version: v2
```

使用 istioctlcreate 创建上述文件，此时默认流量走 Nginx 的 v1，当访问 URL 为/，且 header 的 end-user 是 jason 时访问 v2。

2. 配置延迟

配置延迟为 7 秒，适用于 v1 所有流量：

```
[root@K8S-master01 6.7.1.1]# cat http-Delay.vs.yaml
apiVersion: networking.istio.io/v1alpha3
kind: VirtualService
metadata:
  name: nginx
spec:
  hosts:
  - "nginx.xxx.net"
  gateways:
  - nginx-gateway
#  - mesh
  http:
  - match:
    - uri:
        exact: /
      headers:
        end-user:
          exact: jason
    route:
    - destination:
        host: nginx-svc # name of app service
        subset: v2
  - route:
    - destination:
        host: nginx-svc # name of app service
        subset: v1
        #port:
        #  number: 80
      fault:
        delay:
          fixedDelay: 7s
          percent: 100
```

延迟时间根据业务代码进行设置，通过 istioctlreplace 应用该文件。此时访问 v1（默认流量）会延迟报错，但请求头 end-user 为 jason 时不会报错。

3. 测试故障

验证延迟：

```
[root@K8S-master01 samples]# curl -I nginx.xxx.net
HTTP/1.1 500 Internal Server Error
Server: nginx/1.12.1
Date: Fri, 11 Jan 2019 07:46:35 GMT
Content-Type: text/plain
Content-Length: 18
Connection: keep-alive
```

匹配请求头，将不会出现 500 错误：

```
[root@K8S-master01 samples]# curl -I -Hend-user:jason nginx.xxx.net
HTTP/1.1 200 OK
Server: nginx/1.12.1
Date: Fri, 11 Jan 2019 07:46:50 GMT
Content-Type: text/html
Content-Length: 45
Connection: keep-alive
last-modified: Mon, 11 Jun 2007 18:53:14 GMT
etag: "2d-432a5e4a73a80"
accept-ranges: bytes
x-envoy-upstream-service-time: 7
```

6.7.2 使用 HTTP Abort 触发故障

与 HTTP 延迟解发故障配置不同的是，将 fault 修改为以下内容即可：

```
fault:
    #delay:
    #   fixedDelay: 7s
    #   percent: 100
    abort:
      httpStatus: 500
      percent: 100
```

6.8 Istio 速率限制

为实现速率限制，需要配置 memquota、quota、rule、QuotaSpec 以及 QuotaSpecBinding 这 5 个对象，本示例是官方的 bookinfo。

6.8.1 配置速率限制

（1）配置 Memquota

```
apiVersion: config.istio.io/v1alpha2
```

```yaml
kind: memquota
metadata:
  name: handler
  namespace: istio-system
spec:
  quotas:
  - name: requestcount.quota.istio-system
    maxAmount: 5000
    validDuration: 1s
    overrides:
    - dimensions:
        destination: ratings
        source: reviews
        sourceVersion: v3
      maxAmount: 1
      validDuration: 5s
    - dimensions:
        destination: ratings
      maxAmount: 5
      validDuration: 10s
```

memquota（定义速率限制规则）定义了 3 个不同的速率限制。在没有配合 overrides 的情况下，每秒限制 5000 个请求。

本例 yaml 文件配置了两个 overrides：

- 如果 destination 的值为 ratings，来源为 reviews，并且 sourceVersion 是 v3，即限制速率为 5 秒 1 次。
- 如果 destination 的值为 ratings，无其他匹配条件，即限制速率为每秒 5 次。
- Istio 会选择第一个符合条件的 override 应用到请求上，默认读取顺序为从上到下。

（2）配置 Quota

```yaml
apiVersion: config.istio.io/v1alpha2
kind: quota
metadata:
  name: requestcount
  namespace: istio-system
spec:
  dimensions:
    source: source.labels["app"] | source.service | "unknown"
    sourceVersion: source.labels["version"] | "unknown"
    destination: destination.labels["app"] | destination.service | "unknown"
    destinationVersion: destination.labels["version"] | "unknown"
```

说明：

- quota 模板为 memquota 定义了 4 个 demensions 条目，用于在符合条件的请求上设置 overrides。
- destination 会被设置成为 destination.labels["app"]中的第一个非空的值。
- destination.labels["app"]匹配 app=ratings(memquota 中定义的 destination)的服务
- destinationVersion: destination.labels["version"]匹配版本 version=v1 或者 v2

（3）配置 Rule

```yaml
apiVersion: config.istio.io/v1alpha2
kind: rule
metadata:
  name: quota
  namespace: istio-system
spec:
  actions:
  - handler: handler.memquota
    instances:
    - requestcount.quota
```

Rule 通知 Mixer 使用 instances requestcount.quota 构建对象并传递给上面创建的 handler.memquota。这一过程使用 quota 模板将 dimensions 数据映射给 memquota 进行处理。注意，handler.memquota 和 requestcount.quota 是之前创建的实例。

（4）配置 Quotaspec

```yaml
apiVersion: config.istio.io/v1alpha2
kind: QuotaSpec
metadata:
  name: request-count
  namespace: istio-system
spec:
  rules:
  - quotas:
    - charge: "1"
      quota: requestcount
```

quotaspec 为上面创建的 quota 实例 requestcount 设置了 charge 的值为 1。

（5）配置 QuotaSpecBinding

```yaml
cat quotaspecbinding.yaml
kind: QuotaSpecBinding
metadata:
  name: request-count
  namespace: istio-system
spec:
  quotaSpecs:
  - name: request-count
    namespace: istio-system
  services:
  - name: ratings
    namespace: istio-test
  - name: reviews
    namespace: istio-test
  - name: details
    namespace: istio-test
  - name: productpage
    namespace: istio-test
```

QuotaSpecBinding 把前面的 QuotaSpec 绑定到需要应用限流的服务上，因为 QuotaSpecBinding 所属命名空间和这些服务是不一致的，所以这里必须定义每个服务的命名空间（Namespace）。

使用 istioctlcreate 创建以上文件。

6.8.2 测试速率限制

在浏览器中刷新 productpage 页面。

如果处于登出状态，reviews-v3 服务的限制是每 5 秒 1 次请求。持续刷新页面，会发现每 5 秒钟评级图标只会显示大概 1 次。

如果使用 jason 用户登录，reviews-v2 服务的速率限制是每 10 秒钟 5 次请求。如果持续刷新页面，会发现 10 秒钟之内，评级图标大概只会显示 5 次。

所有其他的服务则会适用于 5000 bps 的默认速率限制。

访问页面显示星星的地方会出现 Ratings service is currently unavailable（评级服务目前不可用）。

要更改 Istio 的其他配置可查看中文官网的网址：https://istio.io/zh/docs/concepts/what-is-istio/。

6.9 小　结

本章介绍了 Istio 的使用，作为 Kubernetes 原生的 Server Mesh，Istio 把业务应用的流量控制、服务发现、安全通信等强大的功能实现变得更加简单、明了。相比于 Spring Cloud 和 Dubbo 等框架，Istio 不仅带来了比这些传统框架更加全面的功能，而且不需要对原有代码进行改动，开发人员也不需要再关注网络层面的实现，开发人员只需要关注业务逻辑即可完成流量控制、服务发现、熔断、负载均衡等功能。同时，Istio 对业务应用和开发人员透明，还可以跨语言使用，大大降低了开发难度，相比于其他框架功能更加强大、配置更加简单，可想而知，Istio 的出现必将带来微服务开发领域的一次颠覆性的变革。